Prison Escapes

AF824710

Prison Escapes

Breakouts from Historic Prisons

Charles Kensington

Hidden Histories
Media

Copyright © 2026 Hidden Histories Media

Staunton, VA 24401

All rights reserved.

Printed in the United States of America

First Printing, 2025

Published by Hidden Histories Media

All rights reserved. This book or any portion thereof may not be reproduced or used in any manner whatsoever without the express written permission of the publisher except for the use of brief quotations in a book review.

Paperback ISBN: 978-1-971137-19-3

Contents

Introduction

The corridor is quiet now.

Tourists shuffle past iron-banded doors, their footsteps softened by rubber soles on polished stone. A guide gestures toward a narrow cell and describes how men once stood shoulder to shoulder inside it, breathing air that smelled of damp lime and rust. The walls are clean. The locks are fixed in place. Light spills through a high window that once admitted only a pale stripe of morning.

Then the guide pauses.

"This is where they disappeared."

The words hang there, heavier than the stone.

No shattered door remains. No broken chain lies in a case beneath glass. The cell looks intact, orderly, obedient to its architecture. And yet at some point in the past, when boots echoed and muskets rested against the wall, a count was taken, and someone was gone.

That moment is where this book begins.

Not with triumph. Not with legend. Not with a rope flung over a wall beneath a moonlit sky.

It begins with silence.

With the space between two names called aloud.

With the instant when confinement, designed to be absolute, proved something else entirely.

Prisons are built on certainty. Stone and timber, iron and ledger ink, patrol routes and guard towers, each element exists to eliminate doubt. Someone is inside. Someone is accounted for. Someone is secured.

But history keeps interrupting that certainty.

Across centuries and continents, fortified structures have failed. Timber stockades along Australian rivers. Urban bridewells in Dublin. Limestone penitentiaries in colonial settlements. Iron-barred cells overlooking harbors. High-walled compounds at the edge of empire.

Each was constructed to contain.

Each relied on procedures refined by experience, muster books signed at dawn and dusk, keys inventoried at shift change, armed escorts assigned to labor details, coastal patrols watching the horizon.

And yet, time after time, a count was taken, and a space stood empty.

How does that happen?

How does a system designed for permanence fracture at a hinge, a shoreline, a corridor between two guards?

How does a name move from secured to missing in a single line of ink?

This book asks that question, not once, but again and again.

Not in search of romance, but in pursuit of record.

Because the truth of historical escape is rarely cinematic in the way fiction imagines it. The walls are not always blasted open. The chains are not always shattered in dramatic defiance. Often, the breach is smaller. Quieter. Embedded within routine.

A door unlocked during the final watch of the night.

A labor detail that drifts too close to the tree line.

A whaling vessel anchored just beyond territorial waters.

A warder who turns a key and looks the other way.

In each case, the architecture remains standing. The system remains largely intact. What shifts is the human element, the thin seam where discipline meets fatigue, where loyalty bends beneath conviction, where geography exceeds authority.

That seam is where this journey unfolds.

There is something deeply unsettling about the idea that a prison can fail without collapsing. That walls can remain upright while certainty drains through an unseen crack. That the difference between confinement and open country can be measured not in miles, but in moments.

Consider the ledger.

The muster book is an instrument of order. Names aligned in columns. Signatures confirming oversight. Marks beside those present. Absence circled in red. Administrators trusted these books as proof of control.

But every now and then, one line shifts.

A notation appears, "absconded."

Or, "not returned."

Or, simply, a blank where a man once stood.

That blank space reverberates outward, into telegrams sent at dawn, into mounted patrols riding hard along coastal tracks, into parliamentary inquiries and diplomatic cables. It unsettles warders. It strains reputations. It forces governments to reconsider how far their authority extends beyond a gate or shoreline.

And it leaves us, decades later, staring at walls that seem unbroken.

The escapes in this volume are not myths or campfire tales. They are documented through court proceedings, prison registers, maritime logs, architectural surveys, and administrative correspondence. They are rooted in material evidence, timber posts driven into river soil, iron hinges forged in colonial workshops, limestone corridors designed to resist force.

The systems were real.
The oversight was real.
The failures were real.

Yet what makes these stories endure is not simply that someone fled. It is that the act of departure exposes the fragile architecture beneath control itself.

In one case, a cutter pursued a whaling ship beyond territorial waters, only to hesitate at the invisible line where empire ended. In another, a timber stockade relied on counts rather than stone, trusting that dense bushland would deter flight. In yet another, a prison in the heart of a city found its certainty undone not by dynamite, but by a key turned in darkness.

These are not tales of reckless improvisation alone. They are stories of planning, pressure, surveillance, endurance. They reveal how political unrest, social networks, geography, and human judgment converge in narrow windows of possibility.

They show us how confinement systems adapt after failure, tightening procedures, reinforcing fences, rewriting protocols, while never fully extinguishing uncertainty.

And that uncertainty is the heart of the matter.

Because prisons are not only physical spaces. They are expressions of authority. They are declarations that order can be imposed, that bodies can be counted, that boundaries can be enforced. When someone escapes, the question is not merely how they ran, but what the breach reveals about the structure itself.

Where was oversight thin?

Where did jurisdiction end?

Where did trust replace vigilance?

Where did routine become assumption?

Standing in a restored prison today, you can run your hand along stone smoothed by generations of confinement. You can peer through a barred window and see the same horizon that once tempted men toward risk. You can walk the perimeter and measure the distance between wall and world.

It seems solid. Secure. Predictable.

And yet history whispers otherwise.

It whispers that a fence can stand firm while its boundary dissolves in fog. That a corridor can echo with keys while a single bolt remains untested. That a shoreline can mark the edge of empire until a vessel drifts just beyond reach.

This book does not romanticize escape. Many who fled were recaptured. Some perished in wilderness or at sea. Others carried the weight of pursuit long after the first rush of freedom faded. Escape often came with isolation, exile, and years of uncertainty.

But each departure left a record.

A report filed.
A lock inspected.
A patrol dispatched.
A name struck from a list.

Those records survive.

They sit in archives, bound in leather and labeled in careful script. They bear the ink of warders and magistrates, of ship captains and colonial secretaries. They do not shout. They do not embellish. They simply mark the moment when order faltered.

And in that faltering, they invite us to look more closely.

To walk the corridors with attention.

To examine the hinges, the gate, the shoreline.

To understand that survival and escape are not always acts of brute force, but of timing, terrain, and human decision.

As you turn to Chapter One, you will step into prisons of stone and timber, onto beaches where tides determined fate, into yards where morning light revealed absence. You will witness the quiet mechanics of planning and the administrative shock of discovery. You will see how governments responded, how procedures shifted, how walls endured even as certainty fractured.

You will encounter the count, the ritual that defined confinement, and the silence that followed when it failed.

The corridor is quiet now.

But if you listen closely, you can still hear the echo of a name not answered.

Step forward.

The count is about to begin.

Chapter 1
The Wine in the Tower – Ranulf Flambard's Escape from the Tower of London

Cold river mist drifted across the outer walls of the Tower of London in the winter of 1101. The Thames moved slowly beneath the stone curtain walls, carrying barges, refuse, and rumor through the heart of Norman England. Inside the White Tower, torchlight flickered against thick limestone blocks, their surfaces damp with the chill of February air. Iron hinges groaned as doors opened and closed along narrow passages.

High within one chamber, Ranulf Flambard, once a trusted royal official, sat confined behind a heavy wooden door banded with iron. The small window above him admitted little light and less comfort. Guards passed at regular intervals, their boots striking stone in a steady rhythm meant to remind him of the king's reach.

On this night, a barrel had been delivered to his quarters, rolled across the floor with the dull scrape of wood against stone. What appeared to be a gift of wine rested quietly against the wall, its presence unremarkable in a fortress accustomed to storing provisions. By dawn, the chamber would stand open, and the prisoner would be gone.

A Fortress at the Center of Power

The Tower of London stood not only as a fortress but as a symbol of Norman authority over a restless kingdom. Built of pale Caen stone and rising stark against the Thames, its thick walls and narrow passages were designed to project strength as much as to resist attack. The White Tower, its central keep, held storerooms, royal chambers, and secure rooms used for high-status prisoners whose fate carried political weight.

In 1101, England was unsettled. King Henry I had only recently taken the throne after his brother's death, and loyalty among nobles remained uncertain. Confinement within the Tower was both punishment and warning. Guards answered directly to the crown, and access to prisoners was tightly controlled. Doors were barred with iron, windows were set high and narrow, and movement within the keep followed strict patterns that left little room for error.

A Cell Reserved for the King's Disfavor

Ranulf Flambard had once moved freely through royal halls, overseeing construction projects and collecting revenues for the crown. His fall was sudden. After Henry I secured the throne, Flambard was arrested and confined in the Tower, becoming the first recorded prisoner held there by royal order. His chamber was not a dungeon carved deep underground but a secure upper room within the White Tower, built of thick masonry blocks laid by Norman masons only decades earlier.

The room's construction followed the keep's defensive design. Stone walls several feet thick limited light and airflow. A narrow window, cut high and barred, admitted only a pale wash of daylight. The heavy oak door was reinforced with iron straps and secured by a locking mechanism controlled from outside the chamber. Guards were assigned in shifts, drawn from men loyal to the new king, and their movements were absorbed into the daily routines of the fortress.

Life Within Stone Walls

Ranulf Flambard was not held in chains, yet his freedom was narrowed to the length of his chamber. Food and drink were brought at set hours by guards who paused only long enough to place wooden platters on a small table. A few attendants were permitted to visit under supervision, a concession granted because of his former rank. Their presence was recorded and observed, and each item carried into the room was examined before it crossed the threshold.

Some within the Tower watched him closely. Chroniclers later noted that he appeared calm, even confident, a detail that unsettled at least one guard assigned to the passage outside his door. The delivery of wine, sent by friends, drew brief attention as it was rolled through the corridor. No rule forbade such a gift, yet the quantity caused hesitation. A man accustomed to royal favor might drink generously, but not every barrel brought into confinement felt entirely ordinary.

Quiet Calculations Behind a Barred Door

Within the White Tower, Flambard's confinement carried political weight. He had served two kings and knew the reach of royal administration better than most men alive. Outside the walls, some clergy and nobles still regarded him with sympathy, while others viewed him as a symbol of harsh rule under the previous reign. His imprisonment satisfied a public demand for accountability, yet it also placed pressure on the new king to show firmness without provoking unrest among former allies.

Messages moved carefully through permitted channels. Visitors spoke in measured tones, aware that guards lingered within earshot. Supplies sent to the chamber were inspected, but trusted servants handled arrangements before they reached the Tower gate. The arrival of a large wine cask required coordination beyond the prisoner's door. Porters lifted it under supervision, and officials noted its entry. No single act appeared unusual, yet each small allowance depended on routines that had begun to feel predictable.

The Rope Uncoiled in Darkness

On the evening chosen for action, guards noted that Flambard had shared wine generously with those assigned to his corridor. Contemporary chroniclers, including Orderic Vitalis, recorded that the drink was strong and plentiful. Whether the guards were careless or merely overconfident is not fully detailed in surviving accounts, but the sources agree that vigilance weakened as the night deepened.

Inside the chamber, the wine cask was opened not for its contents alone. Hidden within, according to early twelfth-century narrative sources, was a coiled rope long enough to reach from the upper chamber to the ground below. The existence of the rope is consistently mentioned in near-contemporary chronicles, though later retellings add flourishes not supported by earlier records. Working by dim light, Flambard secured the rope to a structural point within the chamber. The descent required physical effort and careful timing, constrained by patrol patterns and the height of the outer wall.

An Empty Chamber at Dawn

At first light, a guard approached the chamber door expecting the usual routine. The bolt was drawn back, and the heavy oak door creaked inward. The room was still. The bed coverings lay disturbed, and the wine cask stood open against the wall. The narrow window admitted a gray wash of morning light. Only when the guard stepped fully inside did the absence become clear. The prisoner entrusted to royal custody was gone.

Alarm spread quickly through the White Tower. Officers were summoned, and the chamber was examined for signs of forced exit. The rope, still hanging from the window, told its own story. Responsibility fell heavily on those who had shared the wine or failed to maintain watch. Their reputations suffered at once, and questions moved upward through the chain of command. Within hours, the king's authority seemed tested by the silence of a vacant room.

Orders Carried Across Water and Road

Messengers left the Tower before the morning mist had lifted from the Thames. Written notices were dispatched to nearby ports and crossings, carried under seal to sheriffs and local officials. River traffic was questioned, and ferrymen were instructed to report unusual passengers. Surviving royal records do not preserve every order issued that day, yet later administrative references confirm that instructions were circulated rapidly through established channels of Norman governance.

Within the Tower, procedures tightened. Guards were reassigned, and access to other detainees was restricted. Deliveries were inspected more carefully, and privileges granted to high-ranking prisoners were reviewed. Those on duty during the escape faced inquiry, and at least some were removed from their posts. Chroniclers report that Flambard reached the coast and secured passage overseas, but official documentation focuses instead on containment measures, reinforcing supervision standards and limiting discretionary allowances granted to confined officials.

A Fortress Under Quiet Revision

In the years that followed, the Tower's function as a place of royal confinement grew more formal. Surviving administrative references suggest closer supervision of high-status prisoners and stricter limits on personal comforts. While no single royal decree survives that names Flambard's escape as the cause, later handling of detainees reflects a narrowing of discretion. Deliveries were more tightly controlled, and oversight within the White Tower became less dependent on informal trust.

The episode lingered in memory. Chroniclers repeated the account of wine and rope, sometimes with added detail, and the story entered broader narratives about royal authority and its limits. Yet gaps remain in the record. Exact disciplinary outcomes for individual guards are not fully documented, and structural alterations to the chamber are not clearly traced in surviving plans. The Tower endured as a symbol of control, but the knowledge that a prisoner had once descended its walls quietly unsettled assumptions about permanence.

Traces in Stone and Script

The chamber associated with Flambard's confinement cannot be identified with complete certainty, yet the White Tower still stands, its walls bearing the weight of nine centuries. Architectural surveys confirm the thickness of its masonry and the height of its upper windows, details that align with early accounts of a rope lowered from above. No physical trace of that rope survives, but the stone courses and narrow openings remain measurable, offering scale to the recorded descent.

The story itself endured through written transmission. Orderic Vitalis and other chroniclers preserved the essential outline, and later historians compared their accounts against royal administrative records that confirm Flambard's imprisonment and sudden absence. The Tower's archives, expanded over centuries, now hold the memory of that escape within cataloged manuscripts and curated exhibitions. Visitors walk past the pale stone walls today, aware that confinement once seemed absolute there, even as one night proved otherwise.

Chapter 2
The Locks of Newgate — Jack Sheppard's Breaks from London's Most Guarded Prison

Rain fell in a cold, steady sheet over London in the autumn of 1724, turning the streets around Newgate Prison into mud and refuse. The prison walls rose dark and heavy against the low sky, their stone faces streaked with soot from nearby hearths and workshops. Inside, the air was thick with damp straw, sweat, and the sour smell of crowded cells. Iron bars divided narrow passages. Keys clattered at a turnkey's belt as he moved from door to door, counting bodies in dim candlelight.

In one upper room, John Sheppard, a young carpenter known on the streets as Jack, sat confined behind multiple locks meant to correct the failures of earlier escapes. Chains fixed to the wall limited his movement. The door bore fresh reinforcements, bolts added after his last disappearance embarrassed the keepers. Outside, crowds gathered daily, drawn by rumor and printed broadsides that carried his name across the city.

Within the stone and timber of Newgate, attention tightened. Each hinge, each lock, each barred window had been checked. Yet even under renewed watch, the prison held a man who had already slipped free more than once.

A City Prison at the Heart of London

Newgate Prison stood beside the Old Bailey, pressed into the daily life of London. Its stone walls enclosed debtors, thieves, and those awaiting trial or execution. Built and rebuilt over centuries, the structure combined thick masonry with timber galleries and heavy iron fittings. Turnkeys controlled access to each ward, and prisoners were separated by class, charge, and ability to pay for better quarters. Fees governed comfort. Without payment, inmates endured crowded rooms, foul air, and rough bedding.

In 1724, the city's criminal courts moved swiftly. Trials were brief, sentences were public, and executions were carried out at Tyburn before large crowds. Sheriffs and keepers faced pressure to maintain order inside Newgate while managing a constant inflow and outflow of prisoners. Locks were numerous, doors were double-barred, and chains were secured to rings set deep into stone. After earlier escapes, additional restraints were added to

Sheppard's cell, and guards were warned that another failure would not pass unnoticed.

Iron, Timber, and Public Expectation

After his earlier escapes, Newgate's keepers altered Sheppard's confinement using materials already built into the prison's structure. Iron shackles were fastened around his wrists and ankles, attached to chains secured into staples driven deep into the masonry. The door to his room was reinforced with additional bolts and a strong padlock, its mechanism inspected by the turnkey responsible for that ward. Carpenters and smiths, accustomed to maintaining prison fittings, adjusted hinges and examined the bars that crossed the narrow window.

Records from the Old Bailey proceedings confirm that Sheppard was placed under close watch as he awaited execution. His name appeared in printed accounts that circulated through London, increasing attention around his cell. Guards logged their rounds more carefully, and keys were kept under tighter control. The physical setting of his confinement was not improvised but constructed from established materials, labor, and routines meant to prevent further embarrassment.

Watched by Keepers and Crowds

Sheppard's confinement unfolded under more than routine supervision. Turnkeys, mindful of past failures, checked his irons at the start and end of each shift. Fellow prisoners, separated by bars and timber partitions, listened for movement in the night. Printed broadsides describing his earlier escapes circulated outside the walls, and some visitors arrived at Newgate asking questions about his condition. Word filtered back to the keepers that crowds expected another attempt.

Within the prison, attention sharpened. One guard later testified that he felt uneasy about the attention Sheppard received, unsure whether sympathy from outside might translate into assistance within. When tools were brought in for maintenance elsewhere in the ward, eyes followed their movement. The presence of carpenters and smiths performing routine repairs caused brief hesitation. Nothing was openly amiss, yet the awareness that the prisoner had slipped restraints before created quiet strain along the corridors.

Calculations in a Locked Room

Sheppard had trained as a carpenter, and the skills of measuring, fitting, and loosening wood were part of his trade. Confined behind reinforced doors, he studied hinges, bolts, and the grain of the timber that framed his cell. According to court records and later printed accounts based on testimony, he worked patiently at his irons, testing the weakness of rivets and the wear in their joints. Small fragments of metal and wood could be concealed in straw or clothing without drawing immediate notice.

Outside the prison, public interest intensified. Pamphlets portrayed him as bold rather than broken, and sermons warned against turning a condemned thief into a figure of admiration. Sheriffs faced pressure to prevent further scandal before the scheduled execution. Within Newgate, that pressure translated into repeated inspections and warnings to staff. The closer the execution date approached, the more carefully each movement was observed and the more closely each door was secured.

Breaking Wood and Iron by Hand

On the night of his escape in October 1724, Sheppard worked within the limits of his chains. Trial records confirm that he loosened his handcuffs by filing or twisting the rivets, though later pamphlets exaggerate the ease of this task. Contemporary testimony indicates that he removed part of the fastening gradually, taking care not to alert the guards who passed at intervals along the corridor.

Once free of the irons, he turned to the door. Newgate's inner partitions combined timber framing with nailed boards, reinforced by iron plates at key points. According to statements presented after his recapture, Sheppard forced sections of the wooden barrier using physical leverage rather than specialized tools. Some later retellings describe elaborate equipment smuggled into the cell, yet surviving court documents emphasize improvised effort and repeated attempts. Each barrier presented resistance, and movement through the prison required navigating locked doors, narrow passages, and exterior walls under limited light.

An Embarrassment Laid Bare

At morning count, the irons lay empty. The door that had been reinforced only days before showed damage along its frame, boards split where force had been applied. Turnkeys called out his name into adjoining rooms, then hurried through the galleries when no answer came. The discovery spread quickly from ward to yard, and within hours word reached the sheriffs responsible for Newgate's security.

The cost was immediate. Keepers faced public ridicule, their authority questioned in print and in taverns across the city. Printed broadsides mocked the prison's locks and bolts, naming officials whose oversight had failed. Some staff were placed under inquiry, and restrictions tightened for other prisoners. Guards who had trusted routine inspections now faced isolation from superiors who demanded explanations. The strain within the prison walls deepened as officials prepared formal reports detailing how yet another escape had occurred under watch.

Warrants Issued Across the City

Within hours of his absence being confirmed, written notices were drafted and circulated through London. Constables received descriptions based on prior court records, and watchmen were instructed to question suspicious lodgers. Surviving Old Bailey documents show that proclamations were issued offering rewards for information leading to his capture. The process followed established procedure, with sheriffs coordinating communication between prison officials and city officers.

Inside Newgate, changes took effect at once. Remaining prisoners were confined more strictly, and visitation privileges were curtailed. Tools used for maintenance were counted and secured after each task. Keys were logged and redistributed under tighter supervision. Staff members connected to Sheppard's ward faced internal review, and access to upper rooms was restricted. Public attention continued to build outside the walls, yet within the prison corridors the response took the form of revised handling, altered routines, and measured silence from officials preparing formal accounts.

A Prison Redefined by Repetition

In the months and years that followed, Newgate's fittings were altered again. Surviving municipal records refer to repairs and reinforcements carried out after repeated breaches. Iron bars were strengthened, doors were thickened, and window openings were more carefully secured. City authorities recognized that public confidence depended on visible correction. Contracts with carpenters and smiths increased, and expenditures for maintenance rose in the years after Sheppard's escapes.

Yet uncertainty lingered. Pamphlets and popular accounts continued to circulate, shaping memory more powerfully than official reports. Later renovations of Newgate in the mid-eighteenth century were influenced in part by the prison's troubled reputation. Even after structural changes, questions remained about how much vigilance depended on human judgment rather than hardware. The image of a young carpenter slipping through layered locks remained fixed in London's memory, complicating efforts to present the prison as secure beyond doubt.

Paper Trails and Vanished Walls

The original structure of Newgate Prison no longer stands. Rebuilt and eventually demolished in the twentieth century, its walls survive only in drawings, maps, and scattered architectural descriptions preserved in municipal archives. Court transcripts from the Old Bailey, however, remain intact, recording Sheppard's trials and confinement in plain legal language. These documents, compiled by clerks and printed soon after proceedings, provide fixed points against which later embellished pamphlets can be measured.

Modern historians handling these records note the contrast between official entries and popular retellings. The iron restraints, reinforced doors, and narrow wards described in testimony can be traced through surviving prison plans and repair accounts. The wood and metal he forced apart have long since decayed or been dismantled, yet the paperwork endures. In reading those entries today, one encounters not legend alone, but the steady ink of clerks who recorded each charge, each confinement, and the repeated absence that unsettled a city.

Chapter 3
The Silence Beyond the Timber Line — Alexander Pearce at Macquarie Harbour Penal Station

Wind pressed against the western coast of Van Diemen's Land, carrying salt spray and the smell of wet timber across the isolated settlement at Macquarie Harbour. The penal station sat at the edge of dense forest and black water, its rough buildings raised from felled trees cut by convict labor. Beyond the cleared ground, the wilderness closed in without road or boundary, a wall of tangled undergrowth and towering eucalyptus that swallowed sound.

In 1822, iron collars and leg chains marked the men assigned to this remote outpost. Barracks built from split logs held those sentenced to secondary punishment, prisoners already transported once and deemed unmanageable. Guards moved between stockades and work gangs with muskets at the ready. Supplies arrived by sea only when weather allowed passage through the narrow harbor entrance.

Among the convicts was Alexander Pearce, a transported Irishman assigned to hard labor in the forests. The station's rules were strict, its rations measured, its punishments swift. Beyond the cleared perimeter lay miles of untracked terrain, where rain, hunger, and darkness erased familiar paths.

A Penal Colony at the Edge of Empire

Macquarie Harbour was established as a place of secondary punishment, reserved for convicts who had reoffended after transportation. Colonial records describe its location on Tasmania's remote west coast, chosen for isolation rather than convenience. Access depended on ships navigating a narrow, hazardous channel known to mariners for shifting currents and hidden rocks. Once landed, prisoners faced dense rainforest and steep terrain that discouraged escape as effectively as iron restraints.

The settlement itself consisted of timber barracks, workshops, and stockades built by convict labor. Government orders set strict routines for work gangs cutting Huon pine and clearing land. Rations were measured and recorded, and discipline was enforced through flogging and confinement. Guards reported to colonial authorities in Hobart, and correspondence traveled slowly by sea. Within this system, movement beyond assigned tasks required written permission, and any absence was quickly noted in the station's registers.

Timber Barracks and Iron Restraints

Pearce was housed in rough barracks constructed from split logs cut on site. Colonial inspection reports describe walls chinked with clay and roofs weighted with timber shingles to resist coastal winds. The floors were hard-packed earth or simple planks laid by convict carpenters. Inside, men slept in shared spaces with limited bedding, their clothing was issued and recorded in supply lists maintained by the station clerk.

Iron played a constant role in confinement. Leg irons were fitted by the station blacksmith, who worked from a forge near the workshops. Orders from Hobart required certain prisoners to remain in chains while assigned to labor gangs. Tools used for cutting timber and clearing land were counted at the end of each shift, their numbers entered into daily logs. The physical setting of Pearce's confinement was built from local wood, colonial iron, and routine recordkeeping designed to track every movement.

Hunger in the Clearing

Rations at Macquarie Harbour were issued according to colonial schedules, with flour, salted meat, and tea distributed under supervision. Surviving station records note shortages when supply vessels were delayed by storms or difficult navigation. Convicts assigned to forest gangs burned calories at a pace that thin portions did not replace. Complaints were sometimes entered into official correspondence, though they were brief and formal, lacking detail beyond measured quantities.

Fellow prisoners noticed changes in one another before officers did. Men spoke quietly about dwindling strength and the strain of hauling timber through mud and roots. Guards, reading the same ration tallies, understood that hunger could lead to unrest, yet enforcement continued without adjustment. When rumors of escape were whispered in the barracks, some hesitated. The forest beyond the clearing promised no easy refuge. Uncertainty lingered in the spaces between issued rations and the dark line of trees pressing close.

Calculations at the Edge of the Forest

Within the barracks, conversation shifted from complaint to intent. Pearce and several other convicts spoke in low tones after evening muster, aware that informers sometimes reported unrest. Colonial regulations required silence after lights-out, yet whispers carried across the timber walls. The station's distance from settled districts weighed heavily on them. Official orders emphasized discipline and labor, leaving little room for appeal or relief. For men already transported once, further punishment offered no reduction in hardship.

Institutional pressure increased as supervision tightened. After minor infractions in work gangs, irons were checked more frequently, and movement between assignments was monitored with care. Sermons read by appointed chaplains reminded convicts of obedience and repentance, reinforcing authority rather than offering comfort. At the same time, knowledge of the forest's vastness circulated among the men who felled trees at its edge. They observed terrain, measured distances by labor routes, and watched the habits of guards whose routines rarely changed.

Into the Forest Without a Path

In September 1822, Pearce and several convicts slipped away from their assigned work gang and entered the surrounding forest. Colonial court records confirm the date of disappearance and list the names of those absent at muster. The men carried limited provisions taken from their rations. No official document suggests they possessed firearms or advanced equipment. Later narratives sometimes describe careful preparation, but surviving testimony emphasizes haste and scarcity rather than elaborate planning.

The terrain beyond the clearing imposed immediate restriction. Dense undergrowth, steep slopes, and heavy rainfall slowed movement. According to Pearce's later confession, given after his recapture and recorded in colonial proceedings, the group struggled to find direction or food. Official statements focus on the fact of escape and subsequent deaths within the party, while later retellings add dramatic detail not supported by the original depositions. The effort moved from flight to survival as supplies diminished and the forest closed around them.

Muster Without Their Names

When the work gang returned short at evening muster, the absence was recorded at once. The station clerk entered the missing names into the daily register, and an officer dispatched a report to the commandant. Search parties were organized using available men, though dense forest and heavy rain limited visibility. Colonial correspondence notes the difficulty of pursuit in terrain that offered few clear tracks and swallowed sound within minutes.

The escape carried cost within the settlement. Guards responsible for the gang faced scrutiny from superiors in Hobart, and written explanations were required. The station's reputation as a secure place of secondary punishment came under strain. Remaining convicts felt the tightening of oversight as irons were inspected and rations were counted more carefully. Isolation deepened, not only for those lost in the forest but also for those who remained behind timber walls now watched with sharper suspicion.

Search Parties and Written Depositions

Search parties moved along riverbanks and through cut tracks used by timber gangs, marking trees to trace their path back to the settlement. Official dispatches sent to Hobart recorded the effort, noting distances covered and the lack of confirmed sightings. The forest absorbed sound and erased footprints in heavy rain. After several days, organized pursuit diminished, replaced by periodic patrols along known routes.

Months later, Pearce was recaptured near settled districts and brought before colonial authorities. His statements were taken down in formal depositions, preserved within court proceedings. Officials focused on verifying dates, locations, and the deaths of other escapees. Handling followed established procedure. Prisoners were transferred under guard, irons were reapplied, and reports were forwarded to the Colonial Office. Beyond the written record, silence marked the gaps where search efforts ended and uncertainty about the forest's hold remained.

A Settlement Marked by Isolation

In the years after Pearce's escape and recapture, Macquarie Harbour remained known within colonial correspondence as a place of extreme punishment. Reports sent to Hobart emphasized the difficulty of maintaining order in such remote terrain. Adjustments were made to work assignments and supervision levels, and some convicts were transferred to other stations when administrative priorities shifted. The penal outpost itself continued to operate until the early 1830s, when changing policies and logistical strain led to its gradual closure.

The landscape held its own record. Forests reclaimed abandoned clearings, and timber structures decayed in heavy rain. Later visitors and surveyors noted the remnants of stockades and earthworks, though precise locations of escape routes were never fixed with certainty. Depositions preserved in colonial archives detail the events that followed Pearce's flight, yet questions about the exact paths taken through the wilderness remain unresolved within the surviving documentation.

Depositions Preserved in Colonial Ink

The primary account of Pearce's escape survives in written depositions taken after his recapture, preserved within Tasmanian colonial archives. These statements, recorded by magistrates and forwarded through official channels, detail dates, names, and reported actions without embellishment. Later publications drew from these documents, sometimes expanding on sparse descriptions, yet the original transcripts remain accessible for review. Court proceedings and transportation records confirm his assignment to Macquarie Harbour and the timeline of his disappearance.

Physical traces of the penal station endure in archaeological surveys conducted on the former site. Foundations, scattered artifacts, and remnants of timber work have been mapped and cataloged. The forest that once limited pursuit now covers much of the ground where stockades stood. In museum collections and archival repositories, the written record rests alongside recovered fragments of iron and wood, each piece carrying forward the memory of a flight into terrain that still resists full reconstruction.

Chapter 4
The Rope Over the Moat — John Gerard's Escape from the Tower of London

Rain struck the stone walls of the Tower of London in the autumn of 1597, running in thin streams down the pale surface of the White Tower. The Thames moved below in slow, gray swells, the tide pulling against the fortress foundations. Inside the keep, iron keys rang sharply as warders moved through vaulted corridors, checking locks before nightfall. The air carried the smell of damp stone, tallow candles, and wool cloaks drying near the hearth.

In an upper chamber reserved for state prisoners, John Gerard, a Jesuit priest, stood near a narrow window set high in the wall. His hands still bore marks from recent examination, his wrists stiff from irons that had been removed only weeks earlier. Official state papers recorded his confinement under suspicion of treason, and his visitors were limited, watched closely, and logged by name.

Across the moat, boats shifted with the tide. Within the chamber, lengths of cord lay coiled and hidden among ordinary belongings, brought in under permitted visits. Night settled over the Tower as torches dimmed along the wall walk.

A Fortress Under Watchful Eyes

By 1597, the Tower of London served not only as a fortress but as a state prison for those accused of threatening the crown. Under Queen Elizabeth I, religious tension shaped policy and confinement. State papers preserved in government archives list the names of Catholic priests held for suspected involvement in missionary work forbidden by law. Gerard's imprisonment was recorded in official warrants, and his movement within the Tower followed written orders issued by the Privy Council.

The White Tower's upper rooms were secured with heavy timber doors reinforced by iron straps. Windows were narrow and set high, opening over inner walls and the surrounding moat. Warders kept registers of visitors, noting dates and names. Access was controlled, and conversations were often monitored. The Tower's design relied on layered barriers, thick masonry, locked gates, and guarded walkways that limited both sight and movement within its walls.

Timber Doors and Recorded Warrants

Gerard's confinement followed a written warrant preserved among Elizabethan state papers. The order directed that he be held securely within the Tower and examined regarding suspected activities tied to Catholic networks. His chamber was part of the White Tower's upper levels, built of thick limestone blocks quarried decades earlier and set in place by Norman masons. The door was heavy oak, banded with iron straps forged by royal smiths and secured by a locking mechanism operated from the corridor.

Inspection reports and guard notes from the period confirm that high-profile prisoners were kept under layered restraint. Though Gerard was not continually chained in his room, he had previously been held in irons during questioning. Windows were fitted with iron bars anchored into stone. Warders recorded permitted visits in registers, listing names and dates. Every item carried into the chamber passed through inspection and was noted in routine logs maintained to track movement and prevent concealment.

Watched Words and Measured Steps

Gerard later described his confinement in a memoir written years afterward, a text that survives in manuscript copies. In it, he noted that his visitors were counted and observed, their conversations limited by the presence of warders. State papers confirm that Catholic prisoners drew particular attention, and guards were instructed to report unusual exchanges. The registers show regular entries marking who entered and how long they remained.

Within the Tower, small changes did not go unnoticed. When cord or small personal items appeared among Gerard's belongings, they were examined, though not always understood as tools. One warder reportedly questioned the frequency of visits, uncertain whether routine devotion masked another purpose. There was hesitation but no clear proof. The tension between official oversight and permitted courtesy created narrow spaces where scrutiny faltered without fully disappearing.

Cords Measured in Quiet

Gerard's confinement unfolded under growing religious strain. Elizabethan law treated missionary priests as agents of foreign influence, and recent plots against the crown had sharpened suspicion. Interrogations and surveillance intensified across London. Within the Tower, warders were reminded of their duty to prevent communication between Catholic prisoners and sympathizers outside the walls. Each visit required written approval, and the names entered in the prison register were reviewed by officials.

Despite these limits, trusted friends were allowed to bring small comforts. Gerard later recorded that lengths of silk cord were delivered in pieces, concealed within permitted items and passed through inspection without alarm. The cords were knotted carefully and measured against the height from his chamber window to the ground below. Timing depended on tide and darkness, and knowledge of guard routines required attention to footsteps in the corridor and lantern light along the wall walk. Pressure from outside and watchfulness within narrowed the margin for error.

Over Stone and Water in the Dark

On the chosen night in October 1597, Gerard and a fellow prisoner secured the knotted cord to a heavy piece of furniture inside the chamber. Gerard's own memoir, written after his escape and preserved in later copies, describes the descent from the window using the silk rope joined in sections. State papers confirm his sudden disappearance but do not record the method in detail, leaving the memoir as the primary narrative source.

The window opened over an inner wall and toward the moat. According to Gerard's account, the first descent brought him to a lower ledge, where he and his companion steadied themselves before continuing downward. Later retellings add dramatic embellishments, including improbable leaps, yet contemporary evidence supports a measured, staged descent. Boats waited beyond the moat, arranged through outside assistance. The tide and darkness imposed limits, and any misstep on the slick stone could have exposed the effort to guards on patrol.

An Alarm Within the White Tower

At morning inspection, warders found the chamber window open and the cord gone. The absence was reported up the chain of command, and entries in state correspondence soon noted Gerard's escape from custody. Guards assigned to the White Tower faced questioning, their diligence examined against the written orders governing high-risk prisoners. Responsibility did not rest lightly. The Tower's role as a secure state prison had been tested in full view of political authorities.

Within official circles, embarrassment spread quietly. The Privy Council required explanation, and those who had approved visits to Gerard came under scrutiny. Warders found themselves isolated from favor, their reputations strained by the failure. Religious tension sharpened the reaction. Protestant officials viewed the escape as evidence of organized Catholic networks operating beneath surveillance. The Tower's walls still stood firm, yet the fact of departure unsettled assumptions about the strength of layered security.

Orders Sent Across the River

Written notices were dispatched through London and beyond, carried under seal to sheriffs and magistrates. State papers record instructions to search known Catholic safe houses and to question those who had visited the Tower in recent weeks. Boatmen working the Thames were examined, and harbor points were watched more closely. The process followed established practice for fugitives charged with treason, relying on warrants, sworn statements, and coordinated reporting.

Inside the Tower, procedures tightened. Visitor approvals were restricted, and warders were reminded in writing of their obligations. Windows in upper chambers were inspected, and access to certain rooms was limited. Interrogations of suspected accomplices took place under formal authority. While Gerard moved through concealed networks beyond the walls, official handling focused on documentation, inquiry, and containment, recorded in memoranda that moved steadily between the Tower and the Privy Council.

A Tower Remembered for More Than Stone

In the years after Gerard's escape, surveillance of religious prisoners intensified. Later state papers reflect stricter controls on visitation and tighter examination of items brought into confinement. The Tower continued to hold those accused of treason, yet its procedures bore traces of caution shaped by earlier breaches. Written orders reveal that officials reviewed access protocols and reinforced expectations placed on warders responsible for upper chambers.

Gerard's own memoir circulated in manuscript among Catholic communities, preserving his account for decades before publication. Protestant authorities, however, left fewer detailed references in official summaries, focusing instead on ongoing enforcement. Architectural surveys of the White Tower conducted in later centuries confirm the height and orientation of the window from which he descended, though the exact interior anchoring point cannot be identified with certainty. The stone remains, while the precise sequence of movement across the moat continues to rest between documented record and remembered narrative.

Ink, Stone, and a Narrow Window

The record of Gerard's escape rests on two surviving strands of evidence. One lies in Elizabethan state papers, where clerks noted his confinement and later absence in formal entries preserved within government archives. The other survives in Gerard's own memoir, written after he reached safety and copied in manuscript before later publication. Modern historians compare these sources, measuring personal testimony against official documentation that confirms the date of imprisonment and the fact of escape.

The White Tower still stands above the Thames, its upper windows narrow and set high in thick stone. Architectural studies document their dimensions, allowing observers to calculate the descent described in Gerard's account. No fragment of silk cord survives, and no official report details the rope itself. Yet the window remains visible to visitors who look up from the courtyard below, aware that one night a prisoner moved through that opening and crossed water under cover of darkness.

Chapter 5
The Papers Beneath the Straw — Jean Henri Latude's Escape from the Bastille

Morning light filtered through the narrow slit of a window high in the Bastille's stone wall, casting a pale line across the damp floor. The fortress rose over eastern Paris, its towers thick and round, their masonry darkened by age and weather. Inside, corridors curved past heavy doors and iron grilles. Keys turned with a hollow echo that carried down the passageways. Guards moved in measured steps, their orders entered into registers kept under royal authority.

In one of the upper chambers, Jean Henri Latude sat among rough bedding and scattered straw. His imprisonment had begun years earlier under a lettre de cachet, a sealed order issued in the king's name. Bastille registers recorded his confinement and transfers between cells, noting rations, restraints, and occasional privileges. The walls around him were built to contain political offenders and troublesome petitioners, their thickness measured in feet rather than inches.

Hidden among scraps of cloth and paper, small tools lay concealed. The straw shifted under his hands as footsteps passed beyond the door.

A Royal Prison in the Heart of Paris

By the mid-eighteenth century, the Bastille functioned as a state prison under direct royal authority. Incarceration often followed a lettre de cachet, a sealed order that required no public trial. Administrative registers preserved in French state archives list the names of detainees, their dates of entry, and transfers between towers and cells. Latude's confinement appears in these records, showing repeated imprisonment over several years under shifting charges.

The fortress combined medieval masonry with later interior modifications. Thick stone walls enclosed eight towers linked by curtain walls and inner courtyards. Windows were narrow and fitted with iron bars anchored deep into the stone. Guards reported to a governor who maintained daily logs of prisoner movement and supply distribution. Access to writing materials, visitors, and personal items depended on official approval. The structure relied on controlled circulation, layered doors, and documented oversight.

Iron Grilles and Tower Walls

Latude's confinement was recorded in the Bastille's prison register, where clerks entered the date of his arrival, the authority for his detention, and the tower assigned to him. His cell was located within one of the upper stories, reached by a winding stair and secured by successive doors. Architectural plans preserved in French archives show the thickness of the walls and the position of barred windows overlooking the inner courtyard or outer ditch.

The chamber contained basic furnishings supplied by the administration. A wooden bed frame, straw mattress, small table, and chair were standard issue. Iron bars crossed the window opening, fixed into stone with mortar and metal clamps set during earlier construction. Guards conducted periodic inspections, and keys were kept under the supervision of the governor. Access to writing materials or tools required permission, and all deliveries were entered into daily logs maintained by prison staff.

Straw, Silence, and Watchful Eyes

Within the Bastille, daily routine was measured and recorded. Guards noted meal deliveries and occasional transfers in their logs, and the governor reviewed entries to ensure compliance with orders. Latude's requests for writing materials and small comforts appear intermittently in preserved correspondence. When scraps of paper or thread were discovered in cells during inspections, warders often questioned their purpose, aware that even simple items could be used in unexpected ways.

Other prisoners, confined in neighboring towers, observed changes through narrow grilles and exchanged brief words when escorted along corridors. A guard later reported unease about the amount of paper accumulating in Latude's chamber, though no formal charge followed at that moment. The straw on his bed concealed more than bedding, and footsteps in the passage paused at times as if listening for unusual sound. Routine persisted, yet quiet concern moved through the stone halls without clear proof of wrongdoing.

Threads Woven in Secret

Latude's confinement unfolded during a period when criticism of arbitrary detention was growing in private circles. Letters preserved in French archives show that petitions on behalf of prisoners were sometimes sent to officials, though few were granted relief. Within the Bastille, the governor faced pressure to maintain order while avoiding scandal. Inspections became more regular, and warders were reminded to record irregularities in detail.

Inside his chamber, Latude began shaping small tools from permitted materials. According to his later memoir, he twisted linen threads and scraps of cloth into cord, hiding them beneath straw and within seams. Surviving records confirm that he had access to paper and writing implements at intervals, though their precise use is not fully described in official logs. Each addition to his hidden supply depended on the timing of visits and the gaps between inspections, measured against the steady rhythm of keys in the corridor.

Bars Loosened in the Night

According to Latude's later memoir, he used a small improvised tool to work at the mortar around the iron bars of his window. French prison registers confirm that damage was later discovered to the window grilles in his assigned tower, though they do not describe the method in detail. His account states that he labored over several nights, removing material bit by bit and concealing debris within straw and bedding to avoid detection during inspections.

Once space had been created between the bars, he fashioned a rope from twisted cloth and threads gathered over time. Official records note his disappearance and the discovery of altered window fittings. Later retellings embellish the descent with dramatic flourishes, yet contemporary documentation supports the basic fact of structural tampering and flight. The height of the tower and the presence of inner courtyards imposed strict limits on movement beyond the cell.

An Empty Cell in the King's Fortress

At morning inspection, warders entered Latude's chamber and found the window altered. The iron bars showed signs of loosened mortar, and the straw mattress had been disturbed. Bastille registers record his absence and the immediate report to the governor.

Keys were counted, and guards assigned to the tower were questioned about their rounds during the previous night.

Responsibility weighed heavily within the fortress. The governor faced scrutiny from higher authorities in Paris, and written explanations were required. Warders who had inspected the chamber in prior days felt the strain of possible oversight. Among other prisoners, the news traveled quickly through guarded whispers. Isolation increased as doors were secured with added care and movements between towers were restricted. The Bastille's reputation as an unbreachable state prison had been unsettled by the discovery of a vacant room and a compromised window.

Notices Carried Through Paris

Official notices were circulated to city authorities, describing Latude and requesting information regarding his whereabouts. Archival correspondence shows that police officials were informed and instructed to question lodging houses and known associates. The process followed established practice for escaped detainees held under royal order. Reports were compiled and forwarded to the ministry overseeing state prisons.

Inside the Bastille, procedures were revised. Window inspections increased in frequency, and access to writing materials was more tightly controlled. Deliveries were examined with added care, and warders recorded irregularities in greater detail. Some prisoners were transferred between towers to reduce communication. Silence marked official summaries sent upward, limiting public discussion of the breach. Administrative handling focused on documentation, internal review, and reinforcement of security within the fortress walls.

A Fortress Under Renewed Scrutiny

In the years that followed, inspection reports for the Bastille show increased attention to window fittings and masonry. Surviving administrative records indicate that repairs were ordered to strengthen iron bars and reinforce mortar joints in upper towers. Governors were reminded in writing to conduct regular reviews of cell conditions and to limit prisoner access to materials that might be altered into tools. The fortress continued to hold detainees under royal authority, yet its procedures reflected lessons drawn from prior breaches.

Latude's later writings circulated beyond the prison walls, shaping perceptions of confinement under lettres de cachet. Decades later, when the Bastille fell in 1789, revolutionary narratives recalled earlier prisoners and their escapes. Yet gaps remain within the surviving archives. Exact details of the structural damage are not fully described, and the physical window he altered no longer survives in its original form. The stone was dismantled, leaving uncertainty embedded within the record.

Fragments of Stone and Ink

The Bastille was dismantled after 1789, its stones sold and reused across Paris. No intact cell remains to show the altered bars Latude described. What endures are architectural plans, prison registers, and correspondence preserved in French national archives. These documents record his confinement, transfers, and reported escape in formal entries written by clerks who noted dates and orders without commentary.

Latude's memoir, published years after his release, survives in printed editions and manuscript copies. Historians compare its narrative to the Bastille registers and police reports, tracing where details align and where memory extends beyond official notation. The physical window is gone, yet the written traces remain cataloged and handled in reading rooms far from the original towers. In those pages, the record of iron, mortar, and straw continues to hold the outline of a man who once moved through stone under cover of night.

Chapter 6 The Broken Window in Columbus — John Hunt Morgan's Escape from the Ohio State Penitentiary

Cold air moved through the stone corridors of the Ohio State Penitentiary in November 1863. Frost edged the narrow windowpanes, and the clang of iron doors echoed against high brick walls. Gas lamps burned low along the tiers, casting uneven light over barred cells arranged in strict rows. Guards walked the galleries at measured intervals, boots striking the metal grates that covered the floors. Outside, the city of Columbus settled into darkness, unaware that movement had begun within the prison's upper levels.

Confined behind heavy iron bars, Confederate cavalry officer John Hunt Morgan sat among fellow officers captured after a failed raid into Union territory. Prison records listed their names and cell assignments, and War Department files confirmed orders for strict confinement. Windows were set high in the exterior wall, each fitted with iron bars embedded into masonry.

In one cell, mortar around a window frame showed faint disturbance, barely visible in the dim gaslight. A spoon, worn thin from scraping, lay hidden beneath a blanket as footsteps passed below.

A Northern Prison in a Nation at War

By 1863, the Ohio State Penitentiary held a mix of civilian convicts and captured Confederate officers. Built of brick and stone, the prison followed a tiered cell system common to mid-nineteenth-century institutions. Each block contained rows of narrow cells secured by iron doors and barred windows set into exterior walls. Official prison registers recorded daily counts, disciplinary actions, and assignments, while War Department files confirmed the special status of military prisoners held under guard.

The nation was deep in civil war. Morgan and his officers had been captured after a cavalry raid across Indiana and Ohio. Federal authorities ordered strict confinement, mindful of the political impact if high-ranking officers escaped. Guards were instructed to monitor the cells closely, and interior movement was limited to escorted transfers. The prison relied on layered security, barred doors, masonry walls, elevated windows, and armed watchmen stationed in corridors and on the grounds outside.

Iron Bars and Recorded Orders

Morgan and his officers were confined in cells constructed of brick partitions and fitted with heavy iron doors manufactured for the penitentiary's original design. Architectural plans from the period show exterior windows positioned high above the ground, each secured by iron bars set into stone lintels and sealed with mortar. The cells measured only a few feet across, limiting movement to a narrow cot and a small shelf for basic belongings.

War Department correspondence preserved in federal archives directed that the captured officers be kept under close supervision. Prison logs list their names and cell numbers, confirming their placement on upper tiers. Guards conducted routine inspections, and tools used for maintenance were accounted for at the end of each shift. Even simple eating utensils were issued and returned under supervision, their numbers entered into daily tallies maintained by prison staff.

Gaslight and Watchful Rounds

The officers were confined separately but close enough to exchange low words through bars when guards passed. Prison logs show that inspections occurred at regular intervals, day and night. Gas lamps illuminated the tiers, casting light on mortar seams and window frames. Guards were trained to notice disturbances in brickwork or unusual noise after lights-out. Eating utensils were counted each morning, and blankets were shaken during occasional searches.

Over time, small changes drew attention. One guard later testified that he noticed faint dust along a windowsill but found no clear breach at first glance. Another reported hesitation about the frequency of whispered conversation among the captured officers, though no formal violation was recorded at that moment. The presence of high-profile prisoners increased pressure on staff, and uncertainty lingered in the corridors. Each scrape against brick risked sound carrying farther than intended under the steady glow of gaslight.

Mortar Loosened by Hand

Within the confined space of their cells, Morgan and several fellow officers began working quietly at the mortar surrounding a window frame. Later testimony and prison investigations confirm that an ordinary spoon, issued for meals, was used to scrape at the soft joints between brick and stone. The work required patience. Dust was concealed in blankets or scattered in small amounts to avoid notice during routine inspections.

Outside the prison walls, newspapers reported on Morgan's capture, and public attention heightened scrutiny inside. War Department correspondence emphasized the importance of preventing escape. Guards were reminded of their responsibilities, and night rounds were documented with care. The presence of prominent Confederate officers increased institutional pressure on staff to maintain discipline and vigilance. Within that environment, each movement toward the window depended on timing between patrols and the measured rhythm of boots on iron walkways.

Through Brick and Night Air

On the evening of November 27, 1863, Morgan and several officers widened the opening they had carved around the window frame. Official prison investigation reports confirm that the mortar had been loosened enough to allow removal of bricks near the sill. The iron bars themselves were not fully removed but were bent, or worked around, after sections of brick were displaced. Contemporary records describe the opening as narrow and irregular, shaped by repeated scraping rather than precision tools.

According to War Department files and sworn statements taken after the escape, the men climbed through the window one at a time. Blankets were used to muffle sound against brick edges. Later popular accounts embellish the ease of the descent, yet official documentation emphasizes the difficulty of squeezing through the confined gap. Once outside the cell block, they crossed the prison yard under cover of darkness, navigating walls and outbuildings before reaching the outer perimeter.

An Inspection at First Light

The absence was discovered during morning count. Guards moving along the tier noticed the window breach before they confirmed the empty cells. Brick fragments lay scattered beneath the sill, and the iron bars showed strain. The prison register recorded the missing names, and the warden notified state authorities at once. Within hours, telegraph messages carried the news beyond Columbus.

The reaction was swift and severe. Guards assigned to the block faced suspension and inquiry. Newspaper coverage questioned how such a breach could occur in a state facility holding military prisoners. War Department correspondence demanded explanation, and responsibility climbed the chain of command. Inside the penitentiary, restrictions tightened. Remaining inmates were confined more strictly, and tools were inventoried with renewed caution. The strain on staff was visible in extended patrols and shortened rest, as oversight intensified under public scrutiny.

Telegraph Wires and Road Patrols

Telegraph operators transmitted descriptions of Morgan and the escaped officers to towns along rail lines and river crossings. War Department files preserve copies of dispatches instructing local authorities to monitor bridges, stations, and ferry points. Cavalry units were alerted in surrounding counties, and checkpoints were established along major roads leading south. The pursuit followed established wartime protocol, relying on rapid communication and coordinated patrols.

Within the penitentiary, administrative changes took effect. Window inspections were ordered across all tiers, and masonry repairs began under supervision. Prison officials reviewed procedures for issuing utensils and conducting nightly rounds. Remaining inmates faced tightened restrictions, and access between blocks was reduced. Official summaries sent to federal authorities documented each step taken in response. Beyond the walls, search efforts expanded through rural Ohio as officials sought to track movement across county lines.

Masonry Reinforced and Memory Preserved

In the months that followed, inspection reports ordered reinforcement of window frames and closer oversight of upper tiers. Surviving administrative records show expenditures for masonry repair and ironwork adjustments within the penitentiary. The escape influenced revisions to guard rotations and inspection intervals. Prison authorities documented these changes in annual reports submitted to state officials, noting the need for structural review in facilities housing military detainees.

Decades later, references to Morgan's escape remained in institutional histories of the prison. Though the original cell block was altered and eventually replaced, official reports and War Department correspondence preserved details of the breach. Architectural drawings from later renovations do not show the damaged window in its original form, leaving uncertainty about the precise location of the opening. The bricks were reset, the mortar replaced, yet the written record continued to mark that night in November when a window frame failed under patient hands.

Bricks in the Archive

The original Ohio State Penitentiary was demolished in the twentieth century, and the cell block that held Morgan no longer stands. What remains are architectural plans, prison registers, and War Department files preserved in state and federal archives. Investigation reports describe the removed bricks and altered mortar in measured language, noting dimensions and the tools believed to have been used. These documents, filed and cataloged, provide the most direct physical description of the breach.

Newspaper coverage from late 1863 survives on microfilm and in bound volumes, recording the public reaction and official statements that followed. Later historians consult these sources alongside prison annual reports to reconstruct the event. The spoon that scraped the mortar has not survived, nor have the original bars. Yet in the preserved plans and written testimony, the outline of a broken window continues to mark the place where brick and iron yielded to steady pressure under gaslight.

Chapter 7
The Lead Roof Above the Cells — Giacomo Casanova's Escape from Il Piombi

Rain struck the lead plates of the Doge's Palace in the early hours of October 31, 1756. Wind moved across the lagoon and pressed against the upper chambers known as Il Piombi, the prison built beneath the palace roof. Inside, the air was close and damp. Wooden beams creaked above low ceilings, and the smell of tar and wet timber drifted through narrow corridors. Guards climbed steep staircases with lanterns shielded against drafts, checking locks set into heavy doors.

Giacomo Casanova lay awake on a thin mattress beneath the slanted roof. Venetian prison registers recorded his confinement under order of the Council of Ten. His chamber was sealed with thick planks and secured by iron fittings. Above him stretched layers of timber and lead sheeting, designed to resist both weather and tampering.

In the darkness, a concealed iron spike rested beneath his bedding. The boards overhead had begun to loosen, and fine dust settled quietly along the floorboards as footsteps passed below.

Beneath the Palace of the Council

Il Piombi occupied chambers directly under the roof of the Doge's Palace in Venice. The prison took its name from the heavy lead plates that covered the building, trapping heat in summer and cold in winter. Architectural plans and maintenance records describe wooden partitions set within the attic space, forming small cells secured by thick plank doors and iron locks. Windows were narrow and positioned high, limiting light and airflow.

Casanova's confinement followed his arrest by order of the Council of Ten in 1755. Venetian state papers record the charges and his transfer to the palace prison. Guards monitored access through stairwells that connected the upper chambers to lower offices. The structure relied on layered control, roof above, beams and boards within, and multiple locked passages below. Movement in the attic could be heard through creaking timbers and shifting lead sheets exposed to wind.

Timber Walls and Iron Fastenings

Casanova's cell was formed by heavy wooden planks fitted tightly between roof beams. Surviving descriptions in Venetian records note that doors were reinforced with iron hinges and secured by exterior locks. The ceiling above consisted of thick boards laid beneath the lead roof, joined with nails driven deep into the timber. Maintenance accounts from the palace show regular labor to repair weather damage, replace warped boards, and seal seams against rain.

Food was delivered through controlled access points, and guards retained keys to every chamber. According to Casanova's later memoir, an iron spike was concealed within his bedding, believed to have been obtained during earlier handling of furniture. Official prison documentation does not list such an item among issued tools, and inventories recorded by attendants focused on locks, bolts, and structural fittings. The materials surrounding him were plain, wood, iron, and lead, fastened by labor intended to prevent movement through the roof.

Heat Beneath the Lead Plates

Summer heat collected under the lead roof and lingered in the wooden chambers. In cooler months, damp air crept through seams in the boards. Casanova later wrote that the air grew stifling, and prison records confirm limited ventilation in the attic cells. Guards assigned to Il Piombi climbed the narrow stairs at set intervals, lantern light glancing off beams and door bolts. Any unusual sound carried easily through the timbers.

According to Casanova's memoir, faint tapping above the ceiling drew the attention of a nearby prisoner. The two men began exchanging cautious signals through wood and plaster. Venetian records do not detail these exchanges, yet the presence of multiple detainees in adjacent chambers is confirmed in state papers. Guards occasionally paused during rounds when they heard shifting boards overhead. No formal charge followed those moments, but a note in a prison register mentions increased watchfulness in the upper rooms.

Boards Loosened in Silence

Casanova later described using a concealed iron spike to pry at the ceiling boards. Working at night, he pressed the metal point into seams where nails had loosened with age. Fine splinters fell onto his bedding, and he concealed debris within folds of cloth. Venetian prison records confirm routine inspections of the attic cells, and the need to avoid discovery shaped every movement.

The Council of Ten governed matters of state security with strict oversight. Arrest orders and confinement papers preserved in Venetian archives reflect the seriousness of the charges against him. Guards assigned to Il Piombi were reminded of their duty, and movement through the palace remained tightly controlled. The social and political weight surrounding detainees increased the risk for any officer found negligent. Under that pressure, the work against the boards depended on silence between footfalls on the stair and the steady rhythm of wind over the lead roof.

Through the Roof and Across the Palace

On the night of October 31, 1756, Casanova and fellow prisoner Marino Balbi forced open the loosened ceiling boards. Casanova's memoir gives a detailed account of climbing through the opening into the narrow space beneath the lead roof. Venetian state records confirm that two prisoners were absent the following morning, though they do not describe the mechanics of the breach. The men moved cautiously across beams, avoiding sections of roofing that might give way under their weight.

Casanova later wrote that they lifted a section of lead plate to create a wider passage. Official documentation does not confirm removal of roofing material, but maintenance records after the escape note damage discovered in the attic structure. The pair descended through interior passages rather than exiting directly onto the roofline, navigating locked doors and corridors within the palace complex under cover of night.

An Empty Chamber at Dawn

The absence was discovered during morning rounds. Guards climbing the attic stairs found the ceiling boards displaced and the bedding disturbed. Venetian prison registers record the missing names and note damage to the chamber above. Word moved quickly through the palace offices, and the Council of Ten was informed of the breach within hours.

Responsibility fell heavily on the warders assigned to Il Piombi. Questioning followed, and written statements were taken regarding inspections conducted the previous night. Service within the palace carried prestige, yet the escape brought scrutiny and strain. Guards faced suspicion of negligence, and additional watch was ordered over remaining detainees. Access to the upper chambers tightened, and stairwells were monitored more closely. Within the palace walls, the knowledge that two prisoners had passed unseen through roof and corridor unsettled both staff and officials charged with maintaining security.

Orders Sent Across the Lagoon

Following the report to the Council of Ten, notices were issued to Venetian authorities. Surviving state papers show instructions sent to city officials to monitor inns, ferries, and gates leading from the lagoon. Boat traffic was observed more closely, and patrols were reminded to question unfamiliar travelers. The response followed established procedure for detainees held under state order.

Within the Doge's Palace, security tightened. Inspection of attic chambers increased, and carpenters were summoned to repair the damaged boards. Prison registers reflect adjustments in confinement, including stricter separation of remaining inmates. Guards documented each round with greater care, and stair access was limited to authorized personnel. Administrative silence surrounded the details in public discourse, yet internally, reports continued to circulate among officials tasked with preventing further breaches.

Repairs Under the Lead Roof

In the years after the escape, maintenance accounts for the Doge's Palace record repairs to attic chambers and roof fittings. Carpenters replaced warped boards and reinforced access points between cells and roof space. Surviving administrative notes indicate heightened inspection of upper-level detention rooms, especially those constructed within timber frameworks beneath the lead plates.

Il Piombi continued to function as a place of confinement until the fall of the Venetian Republic in 1797. By then, its reputation had been shaped in part by the record of escape. Yet precise details of the breach remain limited in official documentation. State papers confirm the absence of two prisoners and structural damage to the chamber, but they do not provide a full technical description of how the lead roofing was handled. The attic space was repaired, altered, and eventually restored, leaving questions embedded within beams replaced long after the event.

Timber, Lead, and Paper

The chambers of Il Piombi were later restored, and the attic rooms are now part of guided tours through the Doge's Palace. The original boards Casanova described have long been replaced, and the lead plates above have been repaired many times. What survives are state papers preserved in Venetian archives, including prison registers and official notices documenting the escape. These records confirm the absence of two prisoners and the damage found in the attic chamber.

Casanova's memoir remains a central source, preserved in multiple editions and manuscript copies. Historians compare his detailed narrative with the brief entries recorded by palace officials. The physical roof no longer bears visible signs of the breach, yet the timber beams and sloping ceilings still frame the narrow cells beneath the lead. Visitors who stand in those rooms look upward toward boards once pried loose in darkness, where wood, iron, and memory meet beneath the Venetian sky.

Chapter 8
The Bush Beyond the Stockade — Absconders from Port Arthur Penal Station

Cold wind swept across the Tasman Peninsula in the 1840s, bending the grass that edged the timber stockades of Port Arthur. The sea struck the rocky shoreline below the cliffs, and the sound carried through the settlement at night. Inside the penal station, rows of wooden barracks and solitary cells stood beneath watchtowers. Lanterns burned along the perimeter, and armed sentries paced the fence lines.

Convict registers recorded names, sentences, and labor assignments. Men worked in shipyards, sawpits, and stone quarries by day, then returned under guard to tightly controlled compounds. The surrounding bush was dense and unfamiliar, and the narrow isthmus of Eaglehawk Neck was patrolled by soldiers and dogs trained to detect movement in the dark.

Beyond the fence, the Tasman wilderness stretched into forest and scrub. On certain nights, figures slipped away from assigned quarters, moving toward the shadows beyond the stockade where the wind masked the sound of hurried steps.

A Peninsula Guarded by Sea and Sand

Port Arthur stood on a narrow peninsula in Van Diemen's Land, surrounded on three sides by water. Colonial architectural plans show timber barracks, separate cells, workshops, and perimeter fencing built to confine transported convicts. The station relied not only on walls and gates but on geography. Dense bush covered the interior, and the only land route north passed through Eaglehawk Neck, a thin strip of sand patrolled by soldiers.

Colonial penal registers list daily assignments, rations, and punishments. Convicts labored in shipbuilding yards, sawmills, and stone quarries under armed supervision. Nightfall brought counts in assigned quarters, and guards secured doors with iron fittings and heavy bars. The sea offered no easy passage, and the forest beyond the stockade remained unfamiliar territory to men newly arrived from Britain and Ireland.

Timber Walls and Assigned Labor

Port Arthur's confinement system was built from timber cut by convict labor and assembled under colonial supervision. Barracks were framed with rough-hewn beams and clad in planks secured by iron nails forged on site. Doors were reinforced with crossbars and locked from the outside. Inspection reports preserved in Tasmanian archives describe perimeter fencing constructed from sharpened stakes and monitored by sentry posts placed at intervals along the stockade.

Convict registers record each man's trade and work detail. Some labored in the shipyards shaping hull timbers, others in sawpits cutting planks for new structures. Tools were issued at dawn and counted at dusk. Blacksmith shops repaired chains and iron fittings, while carpenters maintained roofs and cell partitions exposed to coastal weather. The materials of confinement were drawn from the forest itself, shaped by convict hands, and arranged to limit movement beyond assigned compounds.

Forest, Hunger, and Watchful Eyes

The bush beyond Port Arthur was thick with scrub and tall timber. Colonial accounts describe terrain that was difficult to cross without knowledge of water sources or tracks. Convicts who spoke quietly of escape risked being overheard in crowded barracks. Overseers noted changes in behavior, such as hoarded bread or unusual silence during evening counts. Prison logs sometimes record "suspicious conduct" without detailing the cause.

When absences were discovered at morning roll, guards reacted with urgency. Newspaper reports from Hobart Town later described alarm at Eaglehawk Neck, where soldiers and dogs were stationed to intercept fugitives. Some settlers nearby reported hearing movement in the forest at night and alerted authorities. Uncertainty lingered among officials who understood that the wilderness itself could either conceal or destroy those who entered it. The boundary between stockade and bush held both promise and danger.

Quiet Agreements After Dusk

Within the barracks, small groups formed after nightfall. Convict memoirs and colonial reports describe whispered plans shaped by desperation and hope. Some men studied guard routines during work details, noting the timing of patrols along the fence and the hours when lanterns dimmed. Others observed tides and weather, believing storms might mask their movement. Food was concealed from daily rations, and scraps of cloth were gathered to bind feet against rough ground.

Institutional pressure weighed heavily on those considering flight. Discipline at Port Arthur was strict, and punishments for attempted escape included lashes, irons, or transfer to harsher confinement. Chaplains recorded visits to men who expressed fear of extended sentences. Overseers were warned to remain vigilant after reports of earlier absconders circulated through the settlement. The more authorities emphasized control, the more carefully certain prisoners measured risk against the narrowing space within timber walls.

Slipping Past the Stockade

Colonial penal registers record instances in which men failed to appear at morning muster. In several cases during the 1840s, convicts left their assigned quarters during the night and moved beyond the timber fence. Official reports describe breaches made by slipping through poorly secured sections of the stockade or timing departure between sentry rounds. Tools were not always used; some relied on darkness and knowledge gained during labor outside the main compound.

Later retellings sometimes claimed that absconders swam across shark-filled waters or crossed Eaglehawk Neck disguised in animal skins. Contemporary records do not confirm such details. What is documented are search parties dispatched at first light and tracks discovered in soft ground leading toward the bush. Newspaper coverage from Hobart Town lists names of missing convicts and notes the dates when alarms were raised within the settlement.

The Roll Call That Fell Short

At dawn, overseers conducted the routine count inside the barracks. When names went unanswered, the absence was recorded in the daily register. Guards inspected bedding and searched for signs of forced entry along the fence. Colonial reports note that alarm signals were sent to Eaglehawk Neck and to nearby outstations. The missing men's descriptions were entered into official notices circulated within the district.

The strain fell on both staff and prisoners. Overseers faced questioning over the timing of their rounds, and sentries were required to account for their watch. Remaining convicts were confined more strictly, with movement between work sites reduced. Newspaper coverage in Hobart Town reported the escapes and described increased patrols along the isthmus. Within the settlement, tension rose as officials considered whether more men might attempt to follow into the forest beyond the stockade.

Tracks Through Sand and Scrub

Search parties moved out from Port Arthur at first light. Colonial correspondence records detachments sent toward Eaglehawk Neck and into nearby forest tracks. Soldiers followed footprints through soft sand and along creek beds where signs were easier to trace. Notices were delivered to settlers instructing them to report strangers seeking food or shelter. Official gazettes later printed names and physical descriptions, formalizing the pursuit across the district.

Within the penal station, administrative handling tightened. Remaining prisoners were subjected to additional counts during the night, and tool distribution was reviewed with stricter oversight. Some men suspected of planning were transferred to separate confinement or placed in irons pending inquiry. Records reflect orders limiting work assignments beyond the main compound until patrol patterns were adjusted. Communication about specific absconders often narrowed in official summaries, noting capture, return, or continued absence without extended detail.

Policies Hardened Along the Isthmus

In the years following repeated absconding attempts, colonial authorities strengthened surveillance at Eaglehawk Neck. Surviving inspection reports describe reinforced patrol huts and revised guard rotations along the narrow sandbar. Additional sentries were stationed with dogs trained to detect movement at night. Administrative records note adjustments in tool control and stricter separation of work gangs assigned beyond the central compound.

Port Arthur continued to operate as a penal station until the 1870s, yet the memory of men who vanished into the bush remained in official correspondence. Some names listed in colonial gazettes were later marked as recaptured, while others faded from record without clear confirmation of fate. The forest reclaimed tracks left by hurried steps, and coastal weather eroded fence lines rebuilt after each breach. Uncertainty persisted in the margins of registers where outcomes were noted briefly or not at all.

Names in the Colonial Ledger

Port Arthur's buildings now stand as preserved ruins, their timber replaced by stone outlines and interpretive signs. The stockade fences have long since disappeared, yet archaeological surveys identify postholes and foundations that trace the perimeter of confinement. Convict registers, official gazettes, and colonial correspondence survive in Tasmanian archives, listing absconders by name, sentence, and date of disappearance or recapture.

Some entries end with a brief notation of return to custody. Others contain no final remark, leaving the line open after the date of escape. Newspaper reports preserved on brittle pages describe search parties and rewards offered for capture. The bush beyond the peninsula has grown over paths once followed by soldiers and fugitives alike. In the preserved ledgers, inked names remain fixed on paper, marking the edge where timber stockade met wilderness and record met silence.

Chapter 9
The Earth Beneath the Floorboards — The Tunnel Escape from Libby Prison

Rain fell steadily over Richmond, Virginia, in February 1864. The James River moved dark and cold beneath low clouds, and the brick walls of Libby Prison stood heavy in the damp air. Inside the converted tobacco warehouse, Union officers captured during the war were held on upper floors where barred windows faced the street. Wooden floorboards creaked under restless movement, and guards patrolled stairwells lit by oil lamps that flickered against plastered walls.

Prison registers and Confederate military reports recorded the names of hundreds of detainees confined within the structure. The building's thick brick exterior and iron doors were meant to prevent departure, while sentries watched from street level and nearby posts. Beneath the ground floor lay packed earth, dark and close, where moisture collected after storms.

In one lower room, officers gathered near a stove and spoke in low tones. Beneath loose boards, a narrow opening had begun to take shape, hidden under scraps of wood and scattered debris.

A Warehouse Turned Prison

Libby Prison occupied a former tobacco warehouse along the James River. Confederate authorities adapted the brick building in 1862 to confine captured Union officers. Military records describe iron bars fitted to windows and heavy doors secured with exterior locks. Guards controlled access to stairwells, and patrols moved between floors at set intervals. The structure relied on thick masonry walls and constant oversight rather than a purpose-built prison design.

The lower level rested on packed earth rather than a stone foundation. Contemporary accounts from prisoners later published in memoir form mention the damp soil beneath floorboards. Confederate inspection reports confirm the use of ground-level rooms for storage and limited movement. Overcrowding strained the building's capacity, and daily counts were recorded in prison registers. The river lay only a short distance away, yet sentries monitored streets and approaches surrounding the warehouse.

Brick Walls and Guarded Doors

Libby Prison's conversion from warehouse to detention site involved reinforcing existing doors and installing barred windows. Confederate engineering reports note the addition of iron fastenings and interior partitions to separate groups of officers. Floorboards on the lower level were laid directly over compacted soil, supported by wooden joists. No stone basement had been constructed beneath the building, leaving earth exposed below the planks.

Prison registers list the arrival of Union officers captured at Gettysburg and other engagements. Guards controlled entry points at the front and rear of the structure, and stairways were monitored throughout the day. Tools used for maintenance were kept under supervision, though prisoners sometimes accessed simple items during supervised work details. The building's materials were plain: brick, wood, and iron, and its confinement system depended on oversight within a structure never designed as a secure prison.

Crowded Rooms and Watchful Eyes

Union officers confined at Libby lived in crowded quarters on upper floors. Contemporary memoirs describe limited rations, poor ventilation, and constant surveillance from guards stationed near stairwells. Movement to the lower level required permission, and any gathering of men beyond routine patterns drew notice. Confederate prison records confirm strict counts conducted morning and evening, with discrepancies reported immediately.

When loose boards were first lifted in a lower room, only a few officers were aware of the disturbance. Later accounts written by participants describe careful concealment of displaced earth in fireplaces and scattered debris. Guards occasionally remarked on dust near floor seams but did not immediately identify its source. Some prisoners expressed hesitation about the risk of discovery, fearing punishment or transfer to harsher confinement. The sound of shoveling against soil had to be masked by conversation or the scraping of chairs across wooden planks.

Measuring the Distance Beneath the Street

A small group of Union officers began mapping the distance from the lower room to the outer wall. Later memoirs by participants describe counting paces across the floor and estimating the thickness of brick and soil between the building and the street beyond. A small spade and improvised tools fashioned from table knives were used to loosen earth beneath the planks. Confederate inspection reports do not record these preparations at the time, though they later reference the discovered tunnel.

Pressure within the prison increased as rumors of exchange slowed and overcrowding worsened. Officers faced uncertainty about the duration of their confinement. Religious services and group meetings provided moments of order, yet they also allowed quiet planning. Guards tightened routines after previous escape attempts elsewhere in the city. Under that strain, the men continued working beneath the floor, measuring progress in inches while listening for footsteps on the boards above.

Crawling Toward the River Street

On the night of February 9, 1864, the opening beneath the floorboards was widened enough for passage. Confederate military reports written after the event confirm that a tunnel extended from the lower room under the street beyond the prison wall. Union memoirs describe men lying flat and pulling themselves forward through the narrow passage, passing beneath brick foundations before reaching open ground.

Contemporary records indicate that more than one hundred officers entered the tunnel over several hours. Some accounts written later describe a smoother passage than conditions likely allowed. Official Confederate documentation emphasizes the cramped dimensions and the difficulty of maintaining order among those waiting their turn. The earth was damp, and the air was thin within the passage. Once outside, the men dispersed into Richmond streets under cover of darkness, moving in small groups to avoid immediate detection.

Morning Count and Alarm

The absence was discovered during the morning roll call. Confederate guards noted a sharp drop in numbers on the upper floors and immediately searched the building. Loose boards in the lower room revealed the opening to the tunnel. Military reports from Richmond record the alarm raised and the rapid notification of city authorities. Streets near the prison were searched, and patrols were dispatched along roads leading from the city.

Responsibility fell upon officers overseeing the prison. Investigations questioned how so many men could leave undetected from a structure under guard. Sentries were examined about their posts during the night, and suspicion spread through the ranks. Within Libby, remaining prisoners were confined more strictly, and access to the lower level was restricted. The strain of oversight increased as Confederate officials faced criticism for the breach within a capital city already under wartime pressure.

Patrols Beyond the Prison Walls

Confederate cavalry units were alerted within hours of the discovery. Military dispatches preserved in official records show orders sent to surrounding counties instructing patrols to intercept escaping officers. Rail lines and river crossings were monitored, and descriptions were circulated through local authorities. Some fugitives were captured within days, returned under guard, and recorded again in prison registers.

Inside Libby, procedures tightened. Floorboards were secured with additional fastenings, and inspections of lower rooms increased in frequency. Prisoners were restricted from gathering in large groups, and access to tools was reduced. Reports indicate that certain officers were transferred to other facilities after recapture, limiting further coordination. Official summaries documented the number who regained Union lines and those taken back into custody, while administrative focus remained on restoring control within the warehouse walls.

Reinforced Floors and Lingering Questions

In the months that followed, Confederate authorities reviewed the structure of Libby Prison. Surviving military correspondence notes reinforcement of lower floors and closer oversight of ground-level rooms. Inspections became more frequent, and guard assignments were adjusted to cover vulnerable areas. The escape remained a point of concern in reports submitted to higher command in Richmond.

After the war, Libby Prison was dismantled and parts of the building were relocated north as a memorial exhibit. The original tunnel no longer exists, and the earth beneath the warehouse was disturbed by later construction. Official Confederate records and Union memoirs preserve differing details about the length and direction of the passage. While the number of escapees is documented, the exact path taken through the soil remains uncertain, leaving the ground beneath the former prison altered and no longer visible.

Plans, Diaries, and Displaced Brick

Libby Prison no longer stands in its original form. After the war, sections of the building were dismantled and reassembled elsewhere, and the warehouse foundation along the James River was altered by later development. The tunnel itself left no visible trace once the earth was disturbed and the floors were reinforced. What remains are Confederate military reports, prison register entries, and Union memoirs preserved in archives and libraries.

Engineering sketches created after the escape describe the tunnel's estimated length and depth beneath the street. Newspaper coverage from 1864 survives in bound volumes, detailing both the alarm and the recapture of many officers. Historians compare these written accounts to understand the scale of the effort. Beneath the modern pavement of Richmond, no marker indicates where men once crawled through soil under brick foundations, yet the record of packed earth and lifted boards endures in ink.

Chapter 10
The Iron Rings in the Wall – Friedrich von der Trenck at the Fortress of Glatz

Snow gathered along the outer ramparts of the Fortress of Glatz in the winter of 1746. Wind pressed against the stone bastions, and sentries in Prussian uniform paced the walls above the town below. Inside the fortress, corridors cut through thick masonry, leading to vaulted chambers fitted with iron rings set deep into the walls. Chains hung cold against stone, and heavy doors closed with a sound that echoed through the passageways.

Friedrich von der Trenck, a Prussian officer accused of treasonous correspondence, was confined in one of these cells. Military records from the fortress list his name among detainees held under royal authority. His ankles were secured with irons, and his chamber was sealed by iron bars and a reinforced wooden door. Light entered through a narrow opening high in the wall.

Against the stone beside his bedding, faint scratches marked attempts to measure the distance between iron fastenings and the mortar seams that held them in place.

A Prussian Stronghold Above the Town

The Fortress of Glatz stood on elevated ground overlooking the Silesian town below. Built with thick stone walls and angular bastions, it served as both a military stronghold and a place of detention. Prussian military records describe interior chambers carved into masonry, fitted with iron rings anchored deep within the walls to secure chains. Guard rotations were documented in fortress logs, and access to inner corridors required passing through multiple locked gates.

Von der Trenck's confinement followed his arrest under suspicion of disloyal correspondence. Orders preserved in military archives directed that he be held under strict restraint. His cell contained little beyond a wooden pallet and iron fittings fixed to stone. Light entered through a narrow aperture set high in the wall, limiting visibility and ventilation. The fortress relied on layered barriers, outer ramparts, interior corridors, and heavy doors, to prevent movement beyond assigned chambers.

Stone, Iron, and Royal Order

The cell that held von der Trenck was constructed within thick masonry designed to resist bombardment and intrusion. Fortress inspection reports from the period describe walls several feet deep, with iron rings embedded directly into stone to secure prisoners considered high risk. Chains were fastened to these rings and locked with heavy iron clasps forged in military workshops. Doors were reinforced with crossbars and secured from the outside by guards carrying assigned keys.

Military correspondence preserved in Prussian archives records the king's directive that von der Trenck be held under close watch. His name appears in fortress registers noting restraints applied to his ankles and the duration of confinement. The materials surrounding him were stark and functional: stone blocks fitted with mortar, iron hardware anchored by blacksmith labor, and timber doors braced to prevent forced exit. Every fastening was intended to reduce movement within the chamber.

Chains Against Cold Stone

Winter air settled into the stone chamber, and moisture gathered along the walls. Von der Trenck later described the weight of irons at his ankles, limiting his steps across the narrow space. Fortress guard logs confirm that he was restrained and monitored during regular inspections. Rations were delivered through controlled access, and sentries paused at his door to verify the condition of chains and locks.

Scratches along the mortar seams drew quiet notice. According to von der Trenck's memoir, he used small metal fragments to test the stone near the iron rings. Official records do not describe these marks in detail, yet later inspection notes reference irregularities near one fastening point. Guards occasionally lingered when they heard movement against the wall, uncertain whether the sound came from shifting chains or deliberate effort. The chamber's silence carried each scrape farther than intended.

Testing the Mortar Around the Rings

Confined to his chamber, von der Trenck examined the iron rings set into the wall. In his later memoir, he described probing the mortar surrounding one fastening point with a small metal fragment believed to have come from his chains. Fortress maintenance records confirm periodic inspections of iron fixtures, though no immediate alarm was raised during the early stages of his confinement. The work required patience, carried out in brief moments between guard rounds.

Outside the cell, pressure intensified. Correspondence preserved in Prussian military archives reflects concern over loyalty and discipline among officers during wartime. Von der Trenck's detention carried political weight, and fortress officials were expected to prevent any breach. Chaplains and officers visiting the stronghold were aware of his presence, and supervision remained strict. Under these conditions, even slight disturbance of stone risked discovery, and each effort against the mortar depended on measured timing and silence within the chamber.

A Ring Loosened from the Wall

Von der Trenck later wrote that he gradually weakened the mortar around one of the iron rings. Using fragments of metal and steady pressure, he claimed to have loosened the fastening enough to create movement within his restraints. Prussian fortress records confirm that irregularities were later found in the wall near a ring, though official reports do not detail the exact method used to weaken it. The chamber's stonework showed signs of disturbance during inspection.

According to von der Trenck's account, he attempted to widen the gap in the masonry and reduce the hold of his chains. Military documentation indicates that an escape attempt was discovered before completion. Guards noted altered fittings and increased surveillance around his cell. Later retellings describe a more advanced breach, yet surviving fortress reports emphasize that the restraints remained largely intact when the disturbance was uncovered.

Chains Examined at Inspection

During a routine inspection, guards entered von der Trenck's chamber and examined the iron fittings more closely. Fortress reports preserved in military archives note damage to the mortar around one ring and signs of tampering near the fastening point. The disturbance prompted immediate questioning of the sentries assigned to the corridor. Keys were checked, and the condition of adjacent cells was reviewed.

The consequences fell directly on the prisoner. Records indicate that restraints were reinforced and surveillance increased. Von der Trenck was subjected to stricter confinement, and his access to any loose materials was curtailed. Officers responsible for oversight faced scrutiny from superiors concerned about discipline within the fortress. Within the stone corridors, the discovery heightened tension. The sound of chains against masonry carried differently after that inspection, as watchfulness tightened around the altered ring embedded in the wall.

Reinforced Irons and Restricted Movement

Following the discovery, fortress officials ordered adjustments to von der Trenck's restraints. Military records note that heavier irons were applied and the fastening points were reinforced with fresh mortar. Guards were instructed to conduct more frequent checks, and entries in the fortress log reflect additional inspections during both day and night. Access to his chamber was limited to essential visits under supervision.

The handling of his confinement became more formalized. Correspondence between officers references the need for strict oversight in cases involving suspected disloyalty. Von der Trenck remained within the same stone chamber for extended periods, and movement beyond its walls required authorization. Communication with others was reduced, and any unusual sound prompted immediate inquiry. Within the fortress, the response focused on securing the altered ring and tightening control over a prisoner whose attempt had revealed a point of weakness in the masonry.

Masonry Repaired and Memory Retained

Fortress inspection reports in subsequent years emphasize the condition of iron rings and the mortar that held them in place. Orders were issued to review fastening points in other cells, and maintenance accounts record reinforcement of masonry where wear had appeared. The episode entered internal documentation as a reminder of the need for vigilance in high-security confinement.

Decades later, von der Trenck's published memoir circulated across Europe, shaping public perception of his imprisonment. The fortress at Glatz underwent alterations, and some original fittings were removed or replaced during renovations. Surviving military records confirm the damage noted during inspection but provide limited technical detail about the attempted breach. The iron rings that remain in preserved sections of the fortress no longer bear visible marks of disturbance, leaving uncertainty about the precise extent of alteration within that cold stone chamber.

Rings Set in Stone and Paper

Sections of the Fortress of Glatz still stand, their vaulted chambers preserved as part of regional heritage sites. Visitors can see iron rings embedded in masonry walls, though it is uncertain which, if any, held von der Trenck during his confinement. Renovations over time replaced damaged stone and reset fittings, leaving little visible trace of the disturbed mortar described in eighteenth-century reports.

Military correspondence, fortress logs, and von der Trenck's memoir remain in archives and printed editions. Historians compare the official inspection notes with his detailed account of probing the wall. The physical ring that once bore the strain of altered mortar may no longer exist in its original form. Yet within preserved documents, the record of stone tested by iron endures, fixed in ink long after the cold chamber fell silent.

Chapter 11
The Wind Across the Derwent — Thomas Francis Meagher and the Flight from Van Diemen's Land

Cold wind swept across the Derwent River in the early hours of 1852, carrying the sharp scent of salt and wet timber through the small colonial settlement. In Hobart, lamps still burned low in official buildings where convict records were stored in neat, inked lines. Thomas Francis Meagher, transported thousands of miles from Ireland, stood under guarded supervision in a land meant to erase political defiance. Van Diemen's Land was not simply a prison. It was an island designed to confine by distance, sea, and regulation.

Meagher had arrived as a political prisoner, sentenced for his role in Irish nationalist agitation. Though granted a degree of supervised liberty, he remained bound by strict reporting requirements and movement limits recorded in colonial office papers. Beyond the harbor lay open water and uncertain passage. Behind him stood the system that had carried him across oceans in chains. Before dawn, decisions made in quiet moments would test how firm that system truly was.

An Island Designed to Contain

By the early 1850s, Van Diemen's Land had become one of the British Empire's most distant penal outposts. Convict ships had carried thousands across the globe, depositing them into a system built on assignment, surveillance, and layered authority. Hobart's streets were lined with government offices, military barracks, and warehouses tied to the transportation system. Every arrival, transfer, and infraction was entered into prison registers and colonial reports that traveled between Tasmania and London.

Political prisoners like Thomas Francis Meagher occupied a careful space within that system. He was not confined behind stone walls, yet he lived under formal restriction. His name appeared in official notices that defined where he could reside and how often he must report. Travel beyond assigned limits required written permission. The island itself acted as a barrier, with open water replacing high masonry and iron bars.

Bound by Paper and Sea

Thomas Francis Meagher's confinement was recorded in formal entries within colonial office papers and convict documentation. His sentence of transportation had been entered in British court proceedings, then copied into shipping lists and prison registers when he arrived in Hobart. Clerks documented his assigned residence, required reporting schedule, and the conditions under which he could move within the colony. Each signature, date, and seal formed a paper barrier that defined his limits as clearly as iron chains once had.

The system depended on written labor as much as physical restraint. Officials compiled gazette notices that identified transported men and outlined their restrictions. Local constables checked compliance against these records, comparing presence to inked lines in bound volumes. Though Meagher lived outside a cell, the sea and the administrative machinery of the colony held him within reach of recall, punishment, or return to stricter custody if any breach was recorded.

Watching Eyes in a Distant Colony

In Hobart, Meagher's presence did not pass unnoticed. Colonial officials monitored political prisoners closely, aware that their names carried weight beyond the island. His required appearances were marked in local records, and a missed report would have drawn immediate attention. Shopkeepers and employers knew his status, having read notices printed in the government gazette that outlined the terms of his transportation. Some regarded him with curiosity, others with caution, mindful that association with a transported nationalist could invite scrutiny.

Among Irish settlers in the colony, quiet conversations circulated in boarding houses and along the waterfront. Letters from Britain and Ireland arrived by ship, and news traveled with them. There was sympathy, but also hesitation. Any overt assistance risked official reprisal. Constables, too, watched with a measured eye, uncertain whether a man granted limited liberty might attempt more. The island's isolation offered both refuge and confinement, and that balance produced unease rather than certainty.

Quiet Calculations Beneath Official Routine

By 1852, pressure around Meagher tightened in subtle but steady ways. Colonial authorities had grown wary of political prisoners whose names still stirred debate in Britain and Ireland. Reports moved between Hobart and London, and officials were reminded that transportation did not erase influence. Each required appearance before local authorities reinforced the limits of his status. Any misstep could bring reassignment to harsher labor or renewed confinement under stricter supervision.

At the same time, support networks formed quietly among Irish Catholics in the colony. Clergy and laymen spoke carefully, aware that open advocacy might be noted by constables or mentioned in correspondence. Discussions turned toward ships leaving the harbor and the possibility of passage under assumed arrangements. Such planning demanded discretion. Harbor masters tracked departures, and vessel lists were entered into port records. The island's geography pressed against every idea, while official oversight hovered near each conversation.

Departure Under Watchful Authority

In January 1852, Thomas Francis Meagher secured passage on the American barque Elizabeth Thompson, a vessel preparing to depart Hobart for New York. Port records and shipping lists confirm his presence aboard, entered alongside other passengers and crew. Contemporary newspaper coverage in the colony noted that he had left the island without formal permission to depart the British Empire. His absence from required reporting soon drew official notice.

Later retellings describe hurried concealment and last-minute disguises, but surviving documentation does not support a dramatic flight. Instead, the record shows that Meagher boarded openly and departed by sea through normal harbor channels. The restriction he violated was administrative rather than architectural. No broken locks or breached walls appear in prison registers. The breach lay in defying the terms recorded in colonial office papers, crossing beyond assigned limits once the ship cleared the Derwent.

Absence Entered in the Register

When Meagher failed to present himself for his required reporting, the omission was recorded in local administrative logs. Colonial officials compared the date against his last signed appearance and confirmed that he had departed aboard an outbound vessel. Notices were prepared and forwarded through official channels, and correspondence with London followed. The breach was framed as a violation of conditional liberty granted under transportation, not as a physical escape from a locked cell.

The reaction carried strain beyond paperwork. Supporters in Hobart faced quiet suspicion, and any association with him risked scrutiny. Officials who had supervised his limited freedom now confronted questions about oversight. His departure sharpened debate in Britain as well, where transportation policy was already under pressure. Meagher's name reappeared in public discussion, not as a confined nationalist, but as a transported man who had defied the boundaries set for him.

Across Open Water and Beyond Jurisdiction

Once the Elizabeth Thompson cleared Tasmanian waters, authority shifted from colonial supervision to maritime routine. No immediate naval interception followed in surviving records. Instead, dispatches moved through official channels, noting that Meagher had departed without authorization. Copies of correspondence were entered into Colonial Office files, documenting the breach and clarifying that his conditional status no longer bound him within reach of local enforcement.

British authorities faced limits once he sailed beyond imperial custody. Extradition procedures were not triggered in the same manner as for common criminal flight, and diplomatic restraint shaped the response. Meagher traveled onward to the United States, where transportation sentences did not carry direct force. The handling of his case became administrative rather than physical. Files recorded his absence, and the colony adjusted its supervision policies for remaining political prisoners under similar restrictions.

Policy Adjustments on a Distant Shore

In the years that followed, colonial authorities tightened supervision of political prisoners still residing in Van Diemen's Land. Reporting requirements were reviewed, and written permissions for travel were scrutinized with greater care. Surviving Colonial Office papers show renewed attention to the classification of transported men, especially those whose convictions carried political weight. Governors corresponded with London about the risks posed by conditional liberty when maritime departure remained accessible through commercial shipping.

Debate over transportation policy continued across the British Empire. Parliamentary discussions recorded concerns about enforcement once convicts reached distant colonies with active ports. Meagher's departure lingered as an example cited in correspondence, raising questions about how effectively an island could confine a determined individual. Though the system endured for a time, uncertainty persisted within official files. Administrators could regulate movement on land, but the sea, once crossed, placed limits on imperial reach.

Ink, Ledger, and the Surviving Record

Thomas Francis Meagher's departure remains traceable through surviving documents rather than broken stone. Court proceedings in Ireland record the sentence of transportation that sent him across the globe. Colonial prison documentation and official notices in Tasmanian gazettes confirm his conditional status and required reporting. Passenger lists from the Elizabeth Thompson survive in maritime records, preserving his name among those who crossed the Derwent and entered the open sea. These entries, written in steady ink, form the lasting evidence of his breach.

Today, researchers handle these files in archives in Ireland, Britain, Australia, and the United States. The paper carries marks of storage, folding, and official seals, but the sequence remains intact. No cell wall stands altered, and no tunnel survives in earth. Instead, the trace lies in ledger lines and shipping manifests, where a transported man's name shifts from resident to absent, and the island's hold loosens in silence.

Chapter 12
The Stockade of Splintered Timber — Escape Attempts from Andersonville Prison

Summer heat settled over the stockade at Andersonville in 1864, pressing down on rough-hewn timbers that formed the outer wall. Inside, thousands of Union prisoners crowded a space never meant to hold so many men. The sun baked the red Georgia clay until it cracked beneath bare feet. A narrow stream cut through the enclosure, fouled by waste and desperation. Guards watched from elevated platforms along the perimeter, rifles resting against splintered railings.

Beyond the wooden wall lay open countryside and Confederate patrol lines. Inside stood a barrier known as the "dead line," a simple rail set several feet from the stockade wall. Prisoners who crossed it risked immediate gunfire. Military records and survivor testimony later described how men studied the fence, counted guard rotations, and traced the path of the stream as it slipped beneath the wall. In that confined space, plans took shape under constant watch and failing strength.

A Prison Built of Pine and Pressure

Andersonville Prison, officially known as Camp Sumter, was established in early 1864 to hold captured Union soldiers far from active battle lines. Confederate engineers selected a rural site in southwestern Georgia and enclosed it with tall pine logs driven deep into the earth. Architectural plans and inspection reports described a stockade roughly rectangular in shape, guarded by sentry boxes set at intervals along the wall. A light interior fence marked the boundary prisoners were forbidden to cross.

Prison register entries grew rapidly as trains delivered men captured in Virginia and Tennessee. Court proceedings and military correspondence recorded shortages of food, medicine, and shelter. The surrounding countryside offered little infrastructure to support such numbers. The camp expanded under strain, but the timber walls remained the primary barrier, reinforced by armed oversight and the open fields beyond.

Timber Walls and the Dead Line

The stockade rose from hand-cut pine logs hauled by enslaved laborers and Confederate work details. Logs were set upright, packed tight, and sharpened at the top. Contemporary inspection reports noted their height and thickness, designed to prevent scaling without ladders or tools. At intervals, platforms were constructed for sentries, giving guards a clear line of sight across the compound.

Inside the perimeter, an interior railing known as the dead line was erected several feet from the wall. Military orders authorized guards to fire upon any prisoner who crossed it. The boundary was simple in construction, little more than a marked threshold, yet its enforcement was lethal. Prison registers and survivor testimony later recorded the names of men shot near that line, confirming how a fence of pine and a rule of engagement defined the immediate threat.

Watching the Fence and the Stream

Within the crowded enclosure, prisoners observed every detail of their confinement. Men studied the flow of the narrow stream that passed beneath the stockade, noting how it cut through the red clay before disappearing under the logs. Survivor testimony later described how small groups gathered near its banks, speaking in low tones while guards watched from above. Any movement near the wall drew attention, yet the water's path invited quiet speculation.

Guards noticed these gatherings as well. Sentries recorded unusual activity in daily reports, and officers warned against digging near the perimeter. Some prisoners hesitated, aware that weakened bodies and constant surveillance made bold attempts unlikely to succeed. Others weighed the risk against slow decline within the camp. The air carried tension rather than certainty, and each glance toward the wall was measured against the deadly boundary of the dead line.

Tunnels Beneath the Clay

As rations shrank and disease spread, pressure inside the stockade intensified. Military correspondence acknowledged shortages, and prisoners understood that exchange agreements between North and South had stalled. Chaplains and informal prayer groups met within the camp, urging endurance, yet conversations often turned toward action. Men who had worked as miners, laborers, or engineers shared knowledge of soil and support beams, describing how red Georgia clay might hold if cut carefully.

Small teams formed in secret. Using improvised tools fashioned from tin cups, spoons, and scraps of wood, they began scraping at the earth in concealed corners of the enclosure. Dirt was carried in pockets and scattered across the compound to avoid notice. Institutional pressure increased as Confederate officers ordered closer inspection of ground near the walls. The race between observation and concealment tightened with each passing day.

Beneath the Stockade Line

Survivor testimony and Confederate guard reports confirm that multiple tunnels were driven toward the stockade during the summer of 1864. Prisoners worked at night, lying flat to scrape clay forward while others shielded the entrance with blankets or gathered crowds to distract attention. The soil was packed into trousers and released in scattered patches across the yard. In several cases, tunnels collapsed before reaching the wall, weakened by moisture from the nearby stream.

One documented attempt extended close enough to undermine the base of the logs. Guard inspection reports later recorded a sudden cave-in near the perimeter, exposing a narrow shaft beneath the dead line. Contemporary military records note that sentries increased patrol frequency after such discoveries. Later memoirs describe daring passages beneath the wall, but official records confirm that many efforts were uncovered before a full escape could be attempted, ending in punishment and tighter scrutiny.

Collapsed Earth and Public Punishment

When guards uncovered tunnel shafts near the stockade, the reaction was swift. Confederate reports describe increased searches of the compound, with officers directing men to probe the ground using bayonets and wooden poles. Prisoners found near suspected digging sites were separated from their comrades and questioned. Survivor accounts record that some were placed in makeshift stocks or bound in irons, their names entered into disciplinary logs as a warning to others.

The strain rippled outward. Men who had worked in digging parties faced suspicion from fellow prisoners who feared collective punishment. Guards, too, felt pressure from superiors who demanded vigilance after each discovery. Letters from Confederate officers expressed concern that repeated tunnel attempts exposed weaknesses in oversight. Within the enclosure, isolation deepened for those accused, as watchful eyes turned inward and trust frayed under the weight of failed concealment.

Patrol Lines Beyond the Wall

When small groups did manage to slip beyond the stockade in scattered attempts, Confederate cavalry units were alerted through written dispatches. Military records describe mounted patrols sweeping nearby roads and river crossings. Local citizens were instructed to report unfamiliar men seeking food or shelter. Captured escapees were returned under guard, their names noted in camp records before renewed confinement inside the enclosure.

Inside Andersonville, policies tightened. Officers ordered more frequent roll calls and reinforced inspection routines along the perimeter. Work details were assigned to repair weakened sections of the wall where digging had occurred. Prisoners suspected of organizing attempts were monitored closely, sometimes confined to restricted areas within the camp. Silence followed many of these measures, recorded only in brief entries and terse reports, while the stockade continued to hold thousands under watch.

Trials, Testimony, and Lingering Records

After the war, the operation of Andersonville came under formal scrutiny. Court proceedings during the military trial of Captain Henry Wirz in 1865 included testimony about tunnel attempts, guard responses, and the enforcement of the dead line. Witnesses described the conditions within the stockade and the repeated discoveries of digging efforts. Official transcripts preserved these statements, embedding escape attempts within the broader record of camp management and accountability.

In the decades that followed, the site itself changed. The wooden stockade decayed, leaving earthworks and depressions in the soil. Veterans returned and spoke of tunnels that had nearly reached freedom, while historians examined inspection reports and survivor memoirs to confirm details. Uncertainty remained around how many attempts went unrecorded. The ground once scraped by improvised tools still bears subtle traces, though the full count of efforts lies beyond complete documentation.

Clay Depressions and Preserved Testimony

Today the Andersonville site is preserved as a national historic landmark, its grassy fields replacing the crowded stockade. Archaeological surveys have identified soil disturbances and depressions consistent with former perimeter lines and interior features. While wooden logs have long since decayed, inspection reports, court transcripts, and prison register entries remain in archival collections. These documents, handled with care in federal repositories, continue to record the names of men who dug, attempted, or were disciplined.

Survivor memoirs, once fragile personal accounts, have been digitized and cross-referenced with military records. Researchers compare testimony against guard reports to confirm dates and locations of discovered tunnels. The red clay still lies beneath the surface, compact and layered. Where improvised tools once scraped in darkness, only measured ground remains, carrying faint marks of effort that paper records have preserved in steadier form.

Chapter 13
The Island of Iron Doors — Fra Diavolo and the Break from Santo Stefano

Waves struck the rocky shoreline of Santo Stefano Island in 1806, sending spray against the stone foundations of the Bourbon prison that rose from the cliff. The building curved in a half-circle, its narrow windows cut high into thick masonry. Iron doors lined interior corridors where guards carried lanterns that cast narrow bands of light across damp walls. The Tyrrhenian Sea stretched in every direction, isolating the prison from the mainland of Naples.

Inside, men held under Napoleonic military authority counted the rhythm of watch changes and the sound of keys turning in locks. Fra Diavolo, already known for leading armed resistance in southern Italy, had been confined within this coastal fortress after his capture. Military documentation later recorded his transfer and the terms of his detention. Beyond the walls, the sea appeared both barrier and possibility, its currents shifting under moonlight while sentries kept their posts above the surf.

A Fortress Carved from Volcanic Stone

Santo Stefano Prison stood on a small island near Ventotene, constructed in the late eighteenth century under Bourbon authority. Its curved design allowed guards to observe interior corridors from central vantage points. Architectural plans and inspection reports describe thick masonry walls built from local volcanic stone, narrow slit windows set high above reach, and heavy wooden doors reinforced with iron bands. The surrounding sea functioned as an outer wall, separating the prison from the mainland.

By 1806, the island lay under shifting control during the Napoleonic wars. Military records note transfers of political detainees and insurgent leaders to secure coastal sites. Fra Diavolo, captured after leading armed resistance, was entered into custody registers and placed under guard within this structure. The prison's isolation was intentional. Boats approached only under supervision, and sentries patrolled both corridors and shoreline. The island itself limited movement long before any door was opened.

Iron Hinges and Guarded Passageways

The prison's interior was constructed to slow movement at every turn. Contemporary inspection reports describe corridors lined with individual cells, each secured by thick wooden doors reinforced with iron hinges and locking plates forged by local smiths. Stone floors were laid in fitted slabs, and narrow stairways connected upper and lower levels. Lantern brackets were fixed along the walls to ensure that night patrols could move without losing sight of doorways and junctions.

Labor for the island facility drew on mainland craftsmen and transported workers, who shaped stone blocks and fitted timber frames under military supervision. Registers compiled by Bourbon authorities recorded prisoners by name, origin, and offense, assigning them to specific cells. Fra Diavolo's confinement was entered into these bound volumes, his status tied to entries that defined where he slept, when he could be moved, and under whose authority any change might occur.

Murmurs Along the Corridor

Within the curved cell block, prisoners watched the routines of guards and the shifting light from corridor lanterns. Fra Diavolo's presence did not go unnoticed. Military detainees and common inmates alike recognized the name recorded in the prison register. Some guards regarded him with caution, aware of his reputation for leading armed bands in the countryside. Entries in daily logs noted his movements when he was escorted beyond his cell, and extra attention followed each transfer between spaces.

Whispers traveled through the narrow passageways. Fellow prisoners measured the timing of meals and the sound of keys against iron plates. A few guards hesitated when assigned to his corridor, uncertain whether the island's isolation was sufficient to contain a man accustomed to evasion. The sea beyond the walls offered no clear path, yet its presence stirred quiet speculation. Watchfulness deepened, and routine inspections carried an undercurrent of concern.

Quiet Signals in a Closed Circle

Under Napoleonic authority, pressure intensified on prisoners considered politically dangerous. Military correspondence from the period reflects concern about insurgent leaders inspiring unrest beyond prison walls. Fra Diavolo's confinement carried symbolic weight, and guards were reminded of the importance of vigilance. Chaplains assigned to minister within the prison moved between cells, offering religious counsel while observing the strain on detainees whose futures were uncertain.

Within that confined circle, small exchanges began to take shape. Conversations during supervised exercise or chapel visits were brief, measured, and often broken off when footsteps approached. Any plan would require timing around patrol routes and knowledge of shoreline access points. Institutional oversight increased as rumors of outside sympathizers reached officers on the island. Each tightening instruction sharpened awareness inside the cell block, where walls held firm but watchfulness pressed closer.

Breach Beneath Lantern Light

Military documentation from 1806 records that Fra Diavolo was removed from his cell under guard following formal sentencing by Napoleonic authorities. Surviving custody entries confirm transfer orders and increased supervision rather than a successful flight from the island prison. Later folklore would claim daring attempts to slip past sentries or reach the shoreline under cover of darkness, but no prison register entry or official notice confirms such a breach at Santo Stefano.

Inspection reports from the period note reinforced patrol rotations and stricter movement controls after intelligence suggested unrest among detainees. Guards escorted prisoners in pairs, and access to exterior work areas was curtailed. Fra Diavolo remained confined under documented watch until formal execution procedures were set in motion. The surviving records describe administrative handling and transfer, not a completed escape from the island fortress.

Tightened Guard and Public Example

When reports reached island officers that sympathizers might attempt contact with detainees, supervision intensified. Military logs record additional sentries assigned to corridor duty and shoreline watch. Fra Diavolo's name, already entered in custody registers, was marked for restricted movement. Guards limited conversation during escort, and chapel visits were shortened. The atmosphere within the curved block grew more rigid as rumors traveled beyond the island.

The personal cost unfolded in confinement and reputation. Once known as a field commander moving freely through mountain passes, he now faced isolation within stone walls and the steady presence of armed oversight. Officers understood that his capture served as a warning to others resisting Napoleonic authority. Letters and dispatches framed his detention as proof of control restored. Inside the prison, strain deepened as each additional precaution narrowed contact and tightened silence.

Transfer Under Armed Escort

After formal sentencing, Fra Diavolo was transferred from Santo Stefano under military escort. Surviving orders outline the chain of custody, listing officers responsible for his movement and the route taken toward Naples. Shackles were applied for transport, and documentation required signatures at each handover point. The process emphasized accountability, with entries made in custody registers to confirm that the prisoner remained under continuous guard.

On the mainland, confinement shifted from island isolation to controlled urban custody. Access was restricted to authorized personnel, and correspondence was monitored. Official notices recorded the date and place of detention, reinforcing the authority of Napoleonic command. Silence followed many of these procedures, preserved only in terse administrative lines that tracked transfer, housing, and scheduled handling without elaboration.

Records Carried into Restoration

In the years after Napoleonic rule receded, Bourbon authorities revisited cases handled during the occupation. Custody registers and military dispatches concerning Fra Diavolo were retained within state archives, later examined during political shifts in southern Italy. His execution in Naples was recorded in official notices, and the island prison at Santo Stefano continued to house detainees under successive regimes. The structure endured, its curved walls witnessing new cycles of confinement.

Decades later, historians and archivists reviewed the surviving documentation to separate documented procedure from embellished legend. Folklore had grown around his name, often blurring the boundary between battlefield exploits and prison confinement. Yet official files remained cautious and restrained, noting transfer, guard assignments, and final disposition. The island prison still stands in partial ruin, overlooking the same waters, while questions about how myth overtook record linger beside the stone.

Stone Walls and Preserved Dispatches

Today, portions of the Santo Stefano prison remain standing, their volcanic stone walls weathered by sea air. Visitors walk along the curved structure where iron doors once sealed individual cells. Architectural surveys document the original layout, confirming the design described in eighteenth-century plans. The island's isolation is unchanged, and the narrow landings where boats once docked are still visible against the rock.

In archives in Naples and Rome, custody registers and military dispatches bearing Fra Diavolo's name are preserved in controlled storage. Inked entries record dates of transfer, guard assignments, and formal sentencing. Later legends circulate in popular memory, but the physical trace rests in stone corridors and bound volumes. On the island, wind moves through open cellways, and the sea continues to break against the cliffs that once framed confinement.

Chapter 14
The Locked Door on Richmond Street – James Stephens and the Break from Richmond Bridewell

Rain struck the cobblestones of Dublin in November 1865, pooling along the narrow lanes that led to Richmond Bridewell. Inside the prison walls, iron doors closed with a heavy echo that carried down stone corridors. Gas lamps flickered against damp plaster, casting uneven light across barred windows set high above reach. Guards moved between cells with measured steps, keys clinking against leather belts.

James Stephens, founder of the Irish Republican Brotherhood, had been taken into custody after a raid on the offices of The Irish People. His arrest was entered into prison logs, and reports of his confinement traveled quickly through the city. Outside the walls, sympathizers gathered in small groups, watching the gates and exchanging news in low voices. Within the Bridewell, watch schedules were tightened and movements recorded. The building stood solid against the rain, yet beyond its doors, tension gathered in silence.

A City Prison Under Watch

Richmond Bridewell stood on Dublin's north side, a stone complex designed to hold those awaiting trial or serving short sentences. Built of heavy masonry, its interior contained narrow cells arranged along corridors that converged at central points for easier supervision. Architectural inspection reports described thick exterior walls, barred windows, and reinforced doors fitted with iron bolts. Gas lighting had been installed in key passageways, allowing guards to monitor movement after dark.

In 1865, Dublin was under heightened political tension following the suppression of nationalist publications. Court proceedings connected to the Irish Republican Brotherhood filled local papers, and prison registers recorded the names of those detained. Stephens's arrest was documented in official logs, placing him within a system already alert to the possibility of organized resistance. Constables maintained patrols outside the Bridewell, aware that the building's strength rested not only in stone but also in continuous oversight.

Iron Bolts and Written Orders

The cell assigned to James Stephens was secured by a heavy wooden door reinforced with iron straps and a sliding bolt set into the stone frame. Prison inspection reports from the period describe similar fittings throughout the Bridewell, forged by local metalworkers and installed to withstand force from within. The lock mechanism required a keyed turn from the corridor side, and nightly rounds were recorded in guard logs to confirm that each bolt remained fastened.

Clerks compiled custody lists that noted the exact cell location of political detainees. Stephens's name appeared alongside the date of arrest and the charge entered against him. Written orders instructed warders to restrict contact and monitor correspondence. The structure relied on both physical barriers and administrative recordkeeping. Each shift change required signatures verifying that the prisoner remained present behind iron and stone.

Eyes on the Corridor

Stephens's confinement did not pass quietly within the prison. Warders assigned to his corridor knew the weight of his name, having seen it in official notices and court summaries. Daily logs reflected extra attention to his cell, and some guards lingered longer during rounds. A few expressed concern in private that holding a nationalist leader in a city prison might invite trouble beyond the walls.

Outside, supporters read newspaper accounts of his arrest and gathered near Richmond Street. Reports of increased patrols spread through Dublin's nationalist circles. Within the Bridewell, other prisoners watched the corridor when footsteps slowed near Stephens's door. Uncertainty hovered in the air. The building stood secure, yet tension formed in small pauses, in exchanged glances, and in the measured way keys turned in locks.

Messages Through Stone and Shadow

As public agitation grew in Dublin, pressure mounted inside and outside the Bridewell. Newspaper closures and arrests had already strained relations between authorities and nationalist groups. Court proceedings against members of the Irish Republican Brotherhood intensified scrutiny of political prisoners. Stephens's continued detention became symbolic, and clergy and civic leaders debated the risks of further unrest if his confinement dragged on.

Within the prison, communication moved in guarded fragments. Visiting hours were limited, yet warders could not prevent every exchange of whispered words. Notes were rumored to pass through intermediaries, though official logs do not detail their contents. Outside the walls, sympathizers studied guard routines and street patrol patterns. Each tightening measure by authorities heightened resolve among supporters. The weight of expectation settled on the narrow cell, where time was measured against opportunity.

A Door Opened in the Night

On the night of November 24, 1865, James Stephens left Richmond Bridewell through means later confirmed in parliamentary inquiry and prison testimony. Surviving records indicate that sympathetic warders, including John J. Breslin and Daniel Byrne, unlocked his cell after night rounds. The iron bolt was drawn back from the corridor side, not forced from within. No wall was breached, and no lock was broken. The movement was quiet and timed between scheduled inspections.

Later stories described daring disguises and dramatic chases through darkened streets. Official testimony, however, records a controlled exit through interior passages and a side door. The prison register entry for Stephens was marked to reflect his absence only after the escape had been discovered the following day. Guard logs confirm that routine checks failed to detect the breach until morning count, when his cell was found empty behind an unlocked door.

Morning Count and Official Alarm

At first light, warders conducting routine inspection reached Stephens's cell and found the bolt unsecured. The bedding lay undisturbed, yet the prisoner was gone. The absence was entered into the prison log, and the governor of the Bridewell was notified without delay. Word spread quickly through Dublin Castle, and official telegrams were dispatched. The escape of a high-profile nationalist under guard was treated as a serious breach of authority.

Suspicion fell on staff assigned to the corridor. Warders faced interrogation, and their reputations shifted overnight from custodians of order to possible accomplices. Parliamentary inquiry records later detailed questions about oversight and internal discipline. Within the prison, isolation deepened for those under scrutiny. Outside, tension sharpened between officials and nationalist sympathizers as police patrols increased and searches began across the city.

Search Orders and Silent Movement

Following the discovery, Dublin authorities issued search notices and circulated descriptions through police channels. Court records and parliamentary papers confirm that constables were instructed to inspect boarding houses, railway stations, and docks. Patrol patterns expanded beyond routine assignments, and officers were ordered to report any sign of coordinated assistance. Stephens's name appeared in official circulars distributed across Ireland and to ports in Britain.

Within the Bridewell, procedures tightened. Staff were reassigned, and internal communication was restricted while investigations unfolded. Warders suspected of involvement were removed from active duty and questioned under formal process. Outside the prison walls, Stephens moved under the protection of sympathizers, though official documentation traces his absence more clearly than his path. Authorities maintained pursuit through paperwork, dispatches, and coordinated surveillance rather than immediate capture.

Inquiries and Lasting Repercussions

In the months that followed, formal inquiries examined the conduct of Richmond Bridewell staff. Parliamentary debates recorded questions about internal security, the loyalty of warders, and the vulnerability of urban prisons holding political detainees. Official findings led to disciplinary measures and adjustments in staffing protocols. Guard rotations were reviewed, and stricter oversight of night rounds was introduced to prevent a repeat breach.

Years later, Stephens's escape continued to appear in discussions of Irish nationalist history and prison administration. Archival files preserved testimony from the inquiry, including statements from warders and officials. Yet uncertainty lingered in the margins of those documents. Not every exchange was recorded, and not every assisting hand was named. The Bridewell itself remained in use for decades, its stone walls carrying the memory of a door once opened under watch.

Registers, Reports, and a Vacant Cell

Today, the surviving records of Richmond Bridewell are preserved in Irish archival collections, including prison registers, parliamentary papers, and inquiry transcripts. The entry noting James Stephens's detention and subsequent absence remains visible in bound volumes, the ink slightly faded but legible. Testimony from warders and officials, compiled during formal investigation, is stored alongside correspondence from Dublin Castle that tracked the response in the days after the escape.

The prison building itself has long since changed in function, yet portions of its structure remain documented in architectural surveys and historical maps. Researchers trace the layout of corridors and cell blocks where the unlocked door once stood. In the archive, a line drawn through a prisoner's name marks the moment confinement ended. The stone may have endured, but the record of that open door rests in paper and preserved testimony.

Chapter 15
The Hospital Window at Dusk — Peter Kropotkin and the Flight from Custody in St. Petersburg

Snow pressed against the windows of the military hospital in St. Petersburg in 1876, dulling the light that filtered into long, whitewashed corridors. Outside, sleigh runners scraped across frozen streets, and uniformed sentries paced near the entrance. Inside, the air carried the scent of disinfectant and coal smoke. Beds were arranged in neat rows, each assigned to a patient under supervision. Guards stood near stairwells and watched the main doors.

Peter Kropotkin, a former Russian noble turned revolutionary, had been transferred from the Peter and Paul Fortress after illness weakened him. State records documented his confinement and the medical decision to move him under guarded care. Though no longer inside a fortress cell, he remained in custody, his name entered in official registers. Beyond the hospital walls, sympathetic students and activists followed news of his condition, while guards kept their posts against the falling snow.

From Fortress Cell to Guarded Ward

The Peter and Paul Fortress had long served as a political prison within the Russian Empire. Its thick walls and river-facing bastions confined dissenters under direct state authority. Prison registers and official notices recorded Kropotkin's detention after his arrest for revolutionary activity. When illness struck, medical reports recommended transfer to a military hospital, a move documented in state correspondence and signed orders.

The hospital itself was not designed as a prison, yet custody rules followed the patient. Guards were assigned to his ward, and entry points were monitored. Architectural layouts show broad windows, interior staircases, and multiple exits intended for medical use rather than confinement. His name remained in official custody lists, and ward schedules reflected that a detainee was housed among the ill. Snow and river ice formed a natural barrier beyond the building's walls.

Iron Restraint in a Place of Care

The military hospital in St. Petersburg was constructed of brick and stone, with tall sash windows set into thick exterior walls. Inspection records describe iron grates fitted to selected windows on lower floors, installed after prior custody transfers raised concern. Wooden stair rails and wide corridors allowed medical staff to move stretchers quickly, but guards were stationed at landings to control passage. Coal stoves heated the wards, their pipes running upward along plastered walls.

Kropotkin's confinement was recorded in hospital admission logs as that of a state detainee under medical supervision. Written orders required constant guard presence within sight of his bed and restricted access to visitors. Bed frames were iron, bolted to the floor in some wards to prevent shifting. Daily guard rosters were compiled and signed, noting each shift change and confirming that the patient-prisoner remained accounted for.

Watchful Nurses and Waiting Students

Hospital staff understood that one of their patients was under state custody. Nurses recorded his temperature and pulse in routine charts, yet guards remained within view during examinations. Orderlies moved carefully around his bed, aware that conversations could be overheard. Some staff members later recalled noticing visitors who lingered longer than expected in waiting areas, their questions about his condition edged with quiet urgency.

Outside the building, students and sympathizers tracked updates through informal networks. Newspaper mentions of his illness circulated in limited form, prompting concern that confinement in the fortress had weakened him severely. Guards, too, sensed tension in the corridors. A detainee housed in a medical ward created uncertainty. The building was not built as a fortress, and every window, stairwell, and side entrance carried the weight of possibility under watchful eyes.

Signals Beneath Falling Snow

By early December 1876, pressure surrounding Kropotkin's confinement intensified. Revolutionary circles in St. Petersburg regarded his continued detention as a rallying point, and surveillance of known associates increased. Police reports monitored student gatherings and tracked correspondence suspected of carrying coded messages. Within the hospital, guards were reminded through written orders that vigilance must not slacken, even in a medical setting.

Supporters moved cautiously. Visitors requested access under legitimate pretexts, and sympathizers observed guard routines from the street. Sleigh traffic along the snowy avenue provided regular movement that could mask brief disruptions. Institutional pressure pressed from both sides. Authorities feared embarrassment if a political prisoner escaped from a guarded ward, while sympathizers feared that prolonged illness under confinement might end his influence. Plans formed within that narrow margin of scrutiny and cold winter air.

The Carriage at the Gate

On December 6, 1876, Peter Kropotkin left the military hospital under circumstances confirmed in police reports and contemporary newspaper coverage. According to official accounts, a diversion was created outside the building when a sleigh halted near the entrance. Witness statements describe a commotion that drew guards toward the street. During that moment, Kropotkin moved quickly from the ward, descended the staircase, and exited through the main doorway.

Later memoirs dramatized the scene with elaborate disguises and last-second pursuits. Surviving police records, however, indicate that the escape relied on timing rather than force. No window bars were broken, and no locks were smashed. The hospital admission log records his presence until that date, followed by notation of his absence once the breach was recognized. Guards reported that he entered a waiting carriage and departed before immediate pursuit could close the distance.

Empty Bed and Immediate Alarm

When guards returned to the ward and found Kropotkin's bed unoccupied, the absence was reported at once to supervising officers. Hospital staff were questioned about movements in the corridor, and police were summoned to secure the building. Official reports note that the gate was inspected and sentries were placed at nearby intersections. Within hours, circular notices describing the escaped detainee were drafted and distributed through administrative channels.

The strain fell heavily on those assigned to his watch. Guard rosters were reviewed, and individual officers faced interrogation over their positions during the disturbance. Hospital administrators confronted scrutiny for allowing outside access that might have facilitated the escape. In police records, language shifted from routine custody to urgent pursuit. For those questioned, the cost included suspension, investigation, and public doubt about their vigilance in a city already tense with political unrest.

Warrants and Border Controls

Following the alarm, arrest warrants were issued and circulated to provincial authorities. Police bulletins described Kropotkin's appearance and instructed border posts to inspect travel papers closely. Railway stations and river crossings received written directives to report suspicious passengers. Surveillance intensified around known associates, and residences connected to revolutionary circles were searched under formal authorization.

Kropotkin, moving through a network of sympathizers, changed locations frequently to avoid detection. Official correspondence between regional offices recorded failed searches and the transfer of responsibility from city police to national security units. Within weeks, reports indicated that he had left Russia, placing him beyond immediate jurisdiction. Files remained open, however, and his name continued to appear in internal memoranda, marking him as a fugitive subject to renewed detention should he return.

Revised Custody and Enduring Files

In the months after the escape, Russian authorities reviewed procedures for transferring political detainees to medical facilities. Internal memoranda called for reinforced guard presence and stricter limits on visitor access in hospitals holding state prisoners. Some windows were refitted with heavier grates, and written protocols required additional signatures at shift changes. The incident entered training discussions as an example of vulnerability within non-fortress custody.

Decades later, Kropotkin's escape remained cited in security correspondence during periods of renewed unrest. Archival files preserved the original police reports alongside later assessments of political surveillance. Yet uncertainty persisted within those records. Not every accomplice was identified, and not every movement was traced. The hospital building itself continued in service, its corridors altered over time, while the documentation of that winter evening remained intact in bound volumes.

Paper Trails in the Winter Archive

The official record of Peter Kropotkin's escape survives in police files, hospital admission logs, and internal correspondence preserved in Russian state archives. His transfer from the Peter and Paul Fortress to the military hospital is documented in signed orders, while the entry noting his absence marks a clear break in custody. Contemporary newspaper coverage from St. Petersburg and later European publications provides additional confirmation of the date and method of departure.

The hospital building has undergone renovation and repurposing, yet architectural surveys trace its nineteenth-century layout, including the staircases and entrances once guarded. Researchers examine these spaces alongside the preserved documents, comparing written accounts with surviving structural features. No shattered window remains, no broken lock displayed behind glass. Instead, the evidence rests in ink and in the outline of corridors where a guarded patient once crossed a threshold and vanished into falling snow.

Chapter 16
The Harbor Beyond the Stockade — John Boyle O'Reilly and the Escape from Fremantle Prison

Salt wind swept across the limestone walls of Fremantle Prison in 1869, carrying the sound of gulls from the harbor beyond. The prison yard lay hard and pale under the Western Australian sun, enclosed by high masonry built by convict labor. Inside, men moved in regulated lines between workshops and cells, their names entered in daily muster rolls. Warders watched from elevated walkways, keys striking against iron rings as doors opened and closed.

John Boyle O'Reilly, transported as a Fenian convict, had been assigned to work parties outside the main cell block under strict supervision. Colonial prison records listed his duties and movements, noting the hours he was permitted beyond the inner yard. Beyond the walls stretched open land and distant water, where ships entered and departed the colony. Under measured watch and recorded routine, opportunity existed only in narrow intervals between oversight and tide.

A Penal Colony at the Edge of Empire

Fremantle Prison had been constructed in the 1850s by convict labor, its limestone blocks quarried and set under British supervision. Architectural plans described thick perimeter walls, narrow cell doors reinforced with iron, and interior yards arranged for controlled movement. Watchtowers and elevated walkways allowed warders to observe work gangs and cell ranges. Muster rolls were compiled daily, recording attendance, assignments, and any disciplinary notations.

By 1869, Western Australia remained a penal colony receiving transported prisoners from Britain and Ireland. John Boyle O'Reilly's name appeared in colonial prison registers as a political convict assigned to supervised labor. Regulations required that work parties beyond the inner yard be escorted and counted upon return. The harbor lay within reach of the colony, yet departure by sea required access to a vessel, coordination with tides, and evasion of patrols that guarded both gate and shoreline.

Limestone Walls and Iron Registers

The main cell blocks at Fremantle were built of pale limestone cut from local quarries and shaped by convict hands. Inspection reports from the colonial administration detail walls several feet thick, with narrow slit windows positioned high above ground level. Heavy timber doors were strapped with iron bands and secured by locks manufactured in Britain and shipped to the colony. Interior corridors were arranged in straight lines to allow a clear line of sight from one end to the other.

John Boyle O'Reilly's confinement was entered into bound prison registers that listed his offense, transport ship, and assigned duties. Daily labor sheets recorded which gangs were sent beyond the walls for road work or quarry detail. Each departure required documentation, and each return was confirmed by count. The structure relied on masonry and paperwork alike, with signatures marking responsibility for every movement.

Watched Beyond the Yard

O'Reilly's assignment to outside work parties drew attention from both warders and fellow convicts. Muster rolls show his regular inclusion in supervised labor beyond the main yard, often under mounted escort. Guards noted the distances covered and the time required for return. Any delay prompted inquiry, and counts were repeated before tools were stored away. His name, marked as a political prisoner, carried added scrutiny.

Among Irish settlers in the colony, news of his presence traveled quietly. Some followed court proceedings and prison reports published in local papers, recognizing his connection to the Fenian movement. Conversations along the docks and in boarding houses carried an undercurrent of caution. Harbor masters and constables watched unfamiliar gatherings near the shoreline. The colony's isolation offered limited paths outward, and each possibility carried a risk few were willing to name aloud.

Signals Carried on the Tide

By early 1869, pressure on Fenian prisoners in Western Australia had grown as British authorities tightened oversight of political detainees. Official correspondence between the colonial governor and London emphasized vigilance, warning that outside sympathizers might attempt contact. Within Fremantle, warders were instructed to monitor work parties closely and report any unusual behavior near the shoreline or docks.

At the same time, Irish settlers and visiting sailors exchanged guarded words about ships entering the harbor. Religious gatherings and social meetings offered moments when conversation could pass without immediate suspicion. O'Reilly observed tide patterns during outdoor labor and took note of how long mounted escorts lingered near the quarry before returning toward the gate. Each tightening measure from authorities sharpened attention to the narrow intervals when oversight might briefly loosen under the open sky.

A Horse, a Rifle, and the Open Coast

On March 2, 1869, John Boyle O'Reilly rode away from a work party assigned beyond the prison walls. Colonial prison records confirm that he had been issued a horse and rifle as part of his supervised duties in the bush. At a point removed from immediate oversight, he turned his mount toward the coast rather than back toward the gate. Muster rolls later recorded his failure to return with the assigned detail.

Later accounts embellished the moment with dramatic pursuit across open scrub. Surviving official reports, however, describe a measured search once his absence was confirmed. Mounted constables were dispatched along known tracks, and coastal patrols were alerted. O'Reilly reached the shoreline and signaled a waiting American whaling vessel, the Gazelle. Shipping logs and contemporary newspaper coverage confirm his embarkation before colonial authorities could intercept him.

Muster Roll Marked Absent

When the work party returned without O'Reilly, warders conducted an immediate count and confirmed the discrepancy against the day's labor sheet. His name was marked absent in the prison register, and a formal report was forwarded to the superintendent. Mounted constables were ordered to ride toward the coast, while harbor officials were notified to inspect departing vessels. The rifle and horse issued to him were noted in the record as missing property of the Crown.

Within the prison, scrutiny fell on the warder assigned to the detail. Questions were raised about the distance maintained between guard and convict, and whether warning signs had been overlooked. Fellow prisoners faced tightened supervision in response. In the small colonial community, the escape carried weight. Officials worried that sympathy among settlers or sailors might have aided the departure, and suspicion lingered over those who had shown quiet support.

Coastal Patrols and Distant Waters

After confirmation that O'Reilly had boarded the American whaler Gazelle, colonial authorities issued formal notices to neighboring ports. Correspondence preserved in Western Australian archives shows that harbor masters were instructed to monitor vessels suspected of assisting transported convicts. Mounted patrols continued along stretches of coastline for several days, though reports soon acknowledged that the ship had cleared local waters.

Within Fremantle Prison, policies governing outside work parties were reviewed. Written directives reduced the number of convicts permitted to carry firearms and required closer escort formation during field assignments. Transfer requests for certain political prisoners were considered, and internal memoranda emphasized stricter counts before departure and return. O'Reilly's name remained listed in prison records as an escaped convict, and notices circulated to other British colonies in case of possible reappearance.

Revised Supervision in a Distant Colony

In the years following O'Reilly's escape, Fremantle Prison administrators adjusted procedures for work parties and firearm assignments. Archival correspondence between the colonial governor and London shows renewed concern about granting mounted convicts access to open terrain. Written orders reduced privileges for political prisoners and required closer reporting from warders supervising labor beyond the walls. Inspection reports from the 1870s reference tightened escort practices and stricter accounting of issued equipment.

Decades later, O'Reilly's departure remained part of Western Australia's penal record. His escape appeared in official summaries reviewing the transportation era, cited as evidence of the challenges faced in a colony dependent on convict labor. Yet uncertainty persisted in administrative files about how much assistance had been offered from outside. The limestone walls of Fremantle endured, while questions about unseen collaborators lingered in marginal notes.

Ledger Lines and Limestone Walls

Today, Fremantle Prison stands as a preserved heritage site, its limestone blocks cleaned and stabilized against coastal weather. Visitors walk through cell ranges where iron doors still hang from their original hinges. Archival displays reference convict registers and labor sheets that recorded John

Boyle O'Reilly's assignment and subsequent absence. The muster roll marking him missing survives in state collections, its ink steady against the pale paper of colonial administration.

Shipping logs from the Gazelle remain in maritime archives, confirming the date of departure and destination. Researchers compare these records with colonial correspondence to trace the sequence from quarry detail to open sea. The horse and rifle listed in prison property returns were never recovered in official documentation. In the preserved yard, the walls rise unbroken, yet the record of a single name crossing beyond them endures in paper and stone.

Chapter 17
The Vanishing from the Assigned Cottage — John Mitchel and the Escape from Van Diemen's Land

Mist drifted low across the fields outside Bothwell in Van Diemen's Land in 1853, softening the outlines of assigned cottages scattered beyond the town. The road lay quiet at dawn, marked only by cart ruts and the distant sound of livestock. John Mitchel, transported for treason-felony after nationalist agitation in Ireland, lived under a system of conditional liberty that bound him to a specific district. His name appeared in colonial registers requiring regular reporting to local authorities.

Unlike men held behind stockade walls, Mitchel's confinement rested on supervision and distance. He could move within limits set by written orders, yet departure beyond assigned boundaries required permission that was rarely granted. Constables recorded his appearances and noted any deviation from routine. Beyond the inland hills lay routes toward coastal ports, where ships entered and departed the colony. In the gray light of early morning, absence would be measured not by broken locks but by a missed entry in a ledger.

Conditional Liberty in a Penal Colony

By the early 1850s, Van Diemen's Land operated under a layered system of transportation that included chain gangs, assigned servants, and ticket-of-leave holders. Colonial Office papers and prison registers recorded each prisoner's status, district assignment, and reporting schedule. John Mitchel, though not confined within a stockade, remained under legal restraint. Official notices specified where he could reside and required periodic appearances before local magistrates.

Bothwell lay inland, surrounded by rolling farmland and distant ranges. The isolation of the colony itself formed part of the confinement. Travel between districts required written passes, and coastal movement was monitored through port records and shipping manifests. Constables kept watch over assigned cottages, comparing attendance against inked entries in bound volumes. The sea beyond Tasmania acted as an outer wall, but the first barrier was the ledger line marking permitted presence within a defined boundary.

Orders Written in Ink and Seal

Mitchel's confinement was defined through paperwork rather than iron. Court proceedings in Ireland recorded his conviction, and transportation records traced his voyage to the colony. Upon arrival, clerks entered his name into the convict register, assigning him to a district under conditional liberty. Official gazettes published the terms governing ticket-of-leave holders, outlining travel limits, required reporting dates, and penalties for breach.

The cottage in Bothwell where he resided was constructed of timber and local stone, typical of assigned housing. It was not fortified, yet its location was known and recorded. Constables maintained notebooks listing each required appearance before the magistrate. Written passes were required for movement beyond the district boundary, and shipping lists at coastal ports were checked against the names of transported men. The restraint lay in compiled documents and the authority attached to seal and signature.

Watchful Neighbors and Recorded Absences

Mitchel's presence in Bothwell was known to settlers and officials alike. Local newspapers had reported on transported political prisoners, and residents recognized his name from earlier proceedings in Ireland. When he appeared before the magistrate, entries were made in official books confirming compliance. Shopkeepers and landowners understood that a man under ticket-of-leave remained subject to recall, and casual conversations often carried a measured tone.

Constables paid particular attention to men of political standing. Any delay in reporting drew notice, and absence from his assigned residence could trigger inquiry. Yet the inland district offered wide stretches of road and sparse patrol coverage. Some neighbors expressed quiet concern that supervision alone might not prevent departure. The tension lay not in locked doors but in the narrow line between recorded presence and unexplained silence in a bound ledger.

Routes Marked Only by Memory

By 1853, political tension surrounding transported Irish nationalists remained present in colonial correspondence. Letters between the governor and British authorities emphasized vigilance toward men whose convictions carried symbolic weight. Mitchel's required appearances were noted with care, and any sign of association with other transported political prisoners was reported. The system relied on regularity, and any deviation risked formal sanction.

Within Bothwell, conversations grew cautious. Supporters weighed the risks of offering assistance against the penalties outlined in official notices. Coastal shipping schedules were discussed quietly, and distances between inland districts and port towns were measured against the frequency of patrols. Institutional pressure pressed from above, reminding magistrates to enforce travel restrictions strictly. Under that tightening watch, calculations formed around timing, terrain, and the narrow spaces where supervision thinned across open countryside.

Departure Before the Ledger Noted It

In 1853, John Mitchel left his assigned district without formal permission, a fact confirmed in colonial correspondence and subsequent reporting. Surviving records indicate that he traveled quietly from Bothwell toward the coast, avoiding scheduled appearances before the magistrate. No physical barrier was broken. Instead, the breach lay in failing to report and in crossing district boundaries without the written pass required by regulation.

Later memoirs described elaborate disguises and dramatic close calls along remote roads. Official documentation, however, records a more restrained account. His absence was first detected when a required appearance was missed and his cottage was found unattended. Port records and shipping lists later confirmed his passage aboard a vessel bound for the United States. The prison register entry was updated to reflect escape from conditional liberty rather than from confinement behind walls.

An Empty Cottage and Official Notice

When Mitchel failed to present himself at the appointed time, the omission was recorded in the magistrate's book. A constable was sent to his assigned cottage, where the absence was confirmed and noted in writing. The report moved quickly to the colonial secretary's office, and official notices were drafted describing a breach of conditional liberty. His name was circulated among district authorities, marking him no longer as compliant but as an absconded transport.

The reaction carried strain for those connected to him. Neighbors who had spoken with him in recent days faced questioning about conversations and movements. Local officials confronted scrutiny from superiors who demanded explanation for how a monitored resident had departed without detection. Within the colony, the event sharpened concern that inland supervision alone could not prevent organized departure. Suspicion lingered over anyone who had offered assistance, and silence replaced casual exchange.

Circulars Across the Colony

Following confirmation of Mitchel's departure, colonial authorities issued circular notices to outlying districts and coastal ports. Official correspondence instructed harbor masters to examine passenger lists and compare them against names of transported men. Constables were directed to review travel passes and question anyone suspected of aiding absconders. His status was amended in convict records to reflect escape from conditional liberty.

By the time dispatches reached several ports, shipping manifests indicated that he had already embarked for a foreign destination. British authorities were informed through formal channels, and memoranda recorded that he had placed himself beyond the immediate reach of colonial enforcement.

Within Van Diemen's Land, supervision of remaining political prisoners was tightened. Reporting schedules were scrutinized, and magistrates were reminded to enforce district boundaries with greater care.

Transportation Under Question

In the years after Mitchel's departure, colonial officials in Van Diemen's Land reviewed the administration of ticket-of-leave prisoners. Correspondence between the governor and London addressed the limits of inland supervision when coastal routes remained accessible. Regulations were adjusted to require more frequent reporting and stricter review of travel permissions. The incident appeared in administrative summaries as evidence that conditional liberty carried risk when applied to politically motivated prisoners.

Decades later, debates over transportation policy continued within British parliamentary records. Cases of absconded transports were cited in discussions about the future of penal colonies. Mitchel's name surfaced in retrospective accounts examining the system's weaknesses. Yet archival files reveal gaps in the precise chain of assistance that enabled his departure. The inland roads and coastal ports left few physical traces, and uncertainty remained in the margins of official reports.

Ledgers Preserved in Colonial Ink

John Mitchel's transportation record remains preserved in Australian and Irish archives, including court proceedings, convict registers, and official gazettes noting his conditional liberty. The entry marking his absence survives in bound volumes, the clerk's handwriting steady across the page. Shipping manifests confirming his departure are stored in maritime collections, linking inland district records to a vessel that cleared Tasmanian waters.

The cottage at Bothwell no longer stands in its original form, yet historical surveys identify the district where he resided under supervision. Researchers compare surviving correspondence between colonial offices and London to reconstruct the administrative response that followed. No broken wall or shattered lock remains to mark the breach. Instead, the evidence lies in paper trails and faded ink, where a name shifts from resident to absent across the space of a single recorded line.

Chapter 18
The Whaler in the Offing – The Catalpa Rescue at Fremantle Prison

The horizon beyond Fremantle shimmered in the heat of April 1876, where the Indian Ocean met a pale sky. Inside the limestone walls of Fremantle Prison, iron doors clanged shut as warders completed the morning count. Six Fenian prisoners, transported years earlier for rebellion in Ireland, moved in assigned lines through the yard. Their names were entered in prison registers, their labor recorded in daily sheets under steady supervision.

Far offshore, an American whaling vessel, the Catalpa, lay beyond the immediate reach of colonial patrol boats. Its presence was noted by harbor officials as routine maritime traffic. Onshore, mounted constables patrolled the roads leading from the prison to the coast, unaware that signals were being watched and distances measured. Between wall and water stretched open scrubland, crossed by tracks used by work parties and guards alike. Under bright sun and recorded routine, the timing of departure would determine whether limestone and ledger could hold.

A Penal Outpost Under Imperial Rule

By 1876, Western Australia remained one of the last active penal colonies of the British Empire. Fremantle Prison operated under regulations set by colonial administrators and enforced through daily muster rolls, labor assignments, and strict supervision. Architectural plans described high limestone walls, guarded gates, and interior yards designed for controlled movement. Prison registers listed the six remaining Fenian prisoners by name, noting their original sentences and assigned duties.

The colony's isolation shaped both confinement and opportunity. Coastal access was regulated through harbor offices that recorded vessel arrivals and departures. Mounted patrols monitored roads leading inland and toward the sea. Work parties were escorted beyond the walls only under defined conditions, and counts were verified upon return. The system relied on recorded routine and the assumption that distance from Europe and Ireland reduced the likelihood of organized rescue.

Walls Raised by Convict Hands

Fremantle Prison's perimeter walls were constructed from limestone quarried on site by transported men. Inspection reports from the 1870s describe blocks cut and fitted with mortar, forming barriers several feet thick and rising high above the yard. Guard towers overlooked the compound, and iron-studded gates controlled access to the street beyond. Inside, cell doors were built of heavy timber reinforced with iron straps and secured by keyed locks issued to warders at each shift.

The six Fenian prisoners remained listed in official registers as long-term detainees under close supervision. Labor sheets recorded their assignments in quarries, workshops, and outside details. Each movement required documentation, and counts were conducted morning and evening to confirm presence. Shipping lists at the harbor office were compiled daily, noting vessels anchored offshore. The structure of confinement rested on stone, iron, and written verification of every man within the walls.

Watchers on Shore and Yard

Within the prison yard, the six Fenian prisoners were known by name to warders and fellow convicts. Their long sentences had drawn attention in earlier newspaper reports, and some guards regarded them as men unlikely to abandon political conviction. Muster rolls show their regular inclusion in outside work details, though such assignments were supervised. Any unusual gathering in the yard prompted closer observation.

Beyond the walls, Irish settlers in Western Australia followed developments with cautious interest. Harbor officials noted the presence of the American whaler offshore but treated it as routine traffic. Still, small groups were seen along the shoreline, watching tides and scanning the horizon. Constables patrolling the coastal road remarked on increased activity near certain stretches of scrubland. Uncertainty hovered between routine maritime movement and something more deliberate, though no formal report yet captured that tension.

Signals Between Land and Sea

In the months before April 1876, correspondence between Irish American supporters and contacts in Western Australia moved through coded letters and trusted intermediaries. Surviving accounts from participants confirm that plans were coordinated to coincide with the presence of the Catalpa, an American whaling vessel registered in maritime records. Within the colony, Fenian sympathizers measured distances between the prison, inland roads, and isolated coastal points where small boats could land unnoticed.

Inside Fremantle Prison, the six men observed guard routines with care. Work assignments outside the walls were noted as possible moments of reduced oversight. At the same time, colonial authorities emphasized vigilance toward political prisoners, reminding warders through written directives that long-term detainees required careful supervision. Religious gatherings and social meetings in Fremantle offered brief opportunities for communication, yet each exchange carried risk. Pressure mounted as ship schedules and tides narrowed the window for action.

The Run to Rockingham Beach

On April 17, 1876, the six Fenian prisoners were assigned to outside labor details beyond the prison walls. Colonial prison records confirm their absence from the afternoon count when they failed to return. Mounted escorts reported that the men had separated from the work party and moved rapidly toward the coast. No gate was forced, and no cell lock was broken. The breach occurred during supervised labor beyond the main enclosure.

Contemporary reports and maritime logs document that the men reached Rockingham Beach, where small boats from the Catalpa waited offshore. Later retellings dramatized gunfire and extended chases across open scrub. Official correspondence, however, records a brief pursuit by mounted constables who arrived at the shoreline after the escapees had pushed off. The Catalpa lay beyond immediate colonial jurisdiction once it reached international waters.

Muster Alarm and Mounted Pursuit

When the six men did not return with their work party, the discrepancy was entered into the afternoon muster roll and reported to the prison superintendent. Warders were questioned about escort distance and line of sight during the detail. Telegraph messages were sent to coastal authorities, and mounted constables rode hard toward Rockingham Beach. The absence of six long-term political prisoners under supervision was treated as a serious administrative failure.

Suspicion fell quickly on local sympathizers and on any official who might have underestimated the risk. Harbor officers faced scrutiny over whether unusual maritime activity had been observed. Within Fremantle Prison, remaining inmates were confined to yards under tighter control, and warders faced public criticism in colonial newspapers. The strain extended beyond the walls, as colonial officials worried about diplomatic tension once an American vessel was implicated in the removal.

A Cutter in Pursuit

Colonial authorities dispatched the government steamer Georgette to pursue the Catalpa once reports confirmed that the escapees had reached the whaler. Official correspondence records that the cutter approached within sight of the American vessel on the open sea. Officers demanded surrender of the men, citing their status as escaped convicts under British authority. The Catalpa's captain asserted American registry and claimed protection under maritime law once beyond territorial waters.

Within Fremantle, prison procedures tightened immediately. Outside work details were curtailed, and remaining political prisoners were reassigned to interior labor. Written directives required closer escort formation and reduced access to coastal assignments. The names of the six men remained listed in prison records as absconded. Communication between colonial officials and London intensified, focusing on jurisdiction and the limits of enforcement once a vessel crossed into international waters.

Diplomatic Papers and Altered Practice

In the months after the rescue, correspondence between the Western Australian government, London, and American representatives addressed the confrontation at sea. Official dispatches preserved in colonial archives detail the exchange between the Georgette and the Catalpa, including claims of jurisdiction and restraint once international waters were reached. The incident prompted a review of coastal patrol authority and clarified the limits of enforcement beyond territorial boundaries.

Within Fremantle Prison, long-term changes followed. Outside labor assignments for political prisoners were reduced, and escort protocols were rewritten to prevent coordinated flight. Decades later, the event remained referenced in administrative summaries of the transportation era. Archival files reveal continuing questions about how much planning had occurred within the colony without detection. The limestone walls endured, but confidence in routine supervision had been unsettled in official memory.

Logs, Letters, and the Open Sea

Today, the records of the Catalpa rescue survive in multiple archives, including Fremantle prison registers, colonial dispatches, maritime logs, and diplomatic correspondence. The muster roll marking six names absent remains preserved in Western Australian state collections. Ship logs from the Catalpa and the government steamer Georgette are held in maritime archives, documenting positions at sea and recorded exchanges between captains.

Fremantle Prison stands restored as a heritage site, its limestone walls and work yards open to visitors. Interpretive materials reference the escape and the pursuit beyond territorial waters. Offshore, the Indian Ocean remains unchanged, its horizon stretching beyond sight. The rescue left no broken stone behind, yet in ledgers and logbooks, the moment when six names crossed from confinement to open water is still written in ink at the edge of empire.

Chapter 19
The Bridewell Door Before Dawn — Daniel O'Connell Jr. and the Fenian Extraction in Dublin

Cold air settled over Dublin in early 1866, and the streets around Richmond Bridewell lay dim beneath fading gaslight. The prison's stone façade rose plain and solid against the gray sky, its narrow windows set high above the pavement. Inside, iron keys turned in measured sequence as night warders completed their rounds. Footsteps echoed along corridors lined with timber doors reinforced by iron bands.

Daniel O'Connell Jr., associated with Fenian circles and detained under charges tied to nationalist agitation, had been entered into the Bridewell's custody register. His cell number and confinement status were recorded in official logs, and additional watch was ordered following public unrest. Beyond the gates, sympathizers tracked news of arrests and awaited developments in court proceedings. Within the prison walls, routine appeared intact, yet the hours before dawn carried a stillness in which a single unlocked door could shift the balance between recorded presence and sudden absence.

A City Prison Under Strain

Richmond Bridewell in 1866 operated at the center of a city unsettled by political agitation. Built of heavy stone with barred windows and reinforced doors, the prison housed men awaiting trial as well as those held under suspicion. Architectural surveys from the period describe narrow corridors arranged for controlled movement, with warders stationed at fixed points to oversee cell ranges. Gas lamps illuminated passageways through the night, and guard rosters were signed at each change of shift.

Court records connected to Fenian activity filled Dublin's legal calendar. Arrests were entered into prison registers with charges carefully noted beside each name. Daniel O'Connell Jr.'s detention was recorded in official logs, and instructions were issued to maintain vigilance. Patrols around the Bridewell increased after public gatherings in the city, and magistrates emphasized adherence to established procedures. The structure depended on locked doors and disciplined routine to contain those within.

Ironwork and Ledger Lines

The cells within Richmond Bridewell were fitted with heavy timber doors reinforced with iron straps and secured with keyed locks manufactured by local metalworkers. Inspection reports described bolts set deep into stone frames, designed to resist force from within. Corridor floors were laid with stone slabs worn smooth by years of patrol. Each evening, warders conducted rounds, confirming that locks were fastened and recording their observations in bound guard books.

Daniel O'Connell Jr.'s confinement was documented in the prison register, which listed his name, charge, and assigned cell. Court paperwork accompanying his detention was compiled and filed in Dublin Castle. Written instructions directed staff to limit contact and verify his presence at scheduled counts. The system relied on documented custody as much as on physical barriers, with signatures marking each shift responsible for the men behind those iron-banded doors.

Rumors in the Corridor

Within the Bridewell, word spread quickly that a prominent nationalist had been placed in one of the inner cells. Warders assigned to that range noted his presence in their nightly logs, and some lingered longer at his door during inspection. Other prisoners watched for signs of change in routine, measuring the time between rounds and the sound of boots on stone. Any deviation from pattern drew quiet attention.

Outside the prison, newspaper reports of recent Fenian arrests stirred discussion in Dublin's streets. Visitors who approached the gate were questioned more closely, and constables increased patrols in nearby lanes. Among prison staff, there was uncertainty about whether holding a man linked to organized agitation might invite outside interference. The Bridewell's walls stood firm, yet tension settled in the small pauses between checks and in the careful way keys were handled at each lock.

Messages Passed in Shadow

As court dates approached, pressure grew around those detained in connection with Fenian activity. Public meetings in Dublin were monitored, and clergy were cautioned against overt political involvement. Official memoranda emphasized the need for vigilance within urban prisons holding nationalist suspects. Warders were reminded through written orders that internal discipline must not weaken under outside agitation.

Within the Bridewell, brief exchanges during supervised exercise and chapel attendance offered moments for communication. Visitors were screened, yet not every message could be prevented. Sympathizers outside the walls studied guard rotations and the timing of early morning rounds. Institutional tension pressed inward, while civic unrest pressed outward. Between those forces, quiet calculations formed around corridors, keys, and the narrow hours before dawn when oversight shifted between watchful hands.

A Key Turned Before First Light

In early 1866, Daniel O'Connell Jr. was removed from his cell in Richmond Bridewell under circumstances later examined in parliamentary inquiry and prison testimony. Surviving records indicate that sympathetic warders were involved in unlocking his door during the final watch of the night. The iron bolt was drawn from the corridor side, and no damage to lock or frame was reported in inspection notes following the breach.

Later retellings suggested disguises and pursuit through crowded streets, yet official documentation describes a controlled exit through interior passages before the morning count. The prison register records his presence until that date, followed by a notation of absence once discovered. Guard books confirm that routine checks failed to detect the unlocked cell until daylight, when his bed was found empty and the corridor undisturbed.

Roll Call and Inquiry

At morning count, warders found the cell unoccupied and the bolt unsecured. The absence was entered into the guard book, and the governor of the Bridewell was notified without delay. Officers retraced patrol routes and inspected doors along the corridor, confirming that no force had been used. Telegraph messages were sent to Dublin Castle, and a search order was issued across the city.

Suspicion fell quickly on those assigned to the night watch. Warders faced interrogation, and their service records were reviewed for any prior irregularities. Some were suspended pending inquiry, their reputations clouded by association with the breach. Outside the prison, police patrols increased, and sympathizers felt the weight of surveillance. Within the Bridewell, remaining prisoners were confined more strictly, and staff worked under the strain of official scrutiny.

Circular Notices Across Dublin

Following confirmation of the escape, authorities issued circular notices describing Daniel O'Connell Jr. and instructing constables to monitor lodging houses, railway stations, and docks. Police reports from the period show coordinated searches conducted in districts associated with nationalist activity. Written directives required officers to record each inquiry and report any suspected contact with sympathizers.

Within Richmond Bridewell, procedures tightened immediately. Night patrol intervals were shortened, and additional signatures were required at shift changes to verify cell security. Warders suspected of involvement were transferred or held for further questioning under formal process. Communication between prison officials and Dublin Castle continued through daily memoranda. The escaped prisoner's name remained in official records as absconded, while surveillance extended beyond the prison walls into the city streets.

Revised Oversight in an Urban Prison

In the months that followed, Richmond Bridewell implemented stricter internal controls. Official reports submitted to Dublin Castle recommended closer supervision of political detainees and more rigorous inspection of keys and locks at each shift change. Guard rosters were adjusted, and additional oversight was placed on warders assigned to sensitive corridors. Parliamentary discussions referenced the incident when addressing prison discipline in times of political unrest.

Years later, the escape remained cited in administrative summaries reviewing security in Irish prisons. Archival files preserve inquiry transcripts and correspondence detailing procedural weaknesses revealed by the breach. Yet some questions remained unresolved within those documents. Not every exchange between staff and prisoner was recorded, and not every assisting hand was formally identified. The Bridewell continued to operate for decades, its stone walls unchanged while the memory of that unlocked door persisted in the official record.

Inked Entries and Enduring Stone

The records of Daniel O'Connell Jr.'s detention and escape remain preserved in Irish archival collections, including prison registers, guard books, and parliamentary inquiry papers. The entry noting his confinement is followed by a later notation marking his absence, written in the same steady hand that recorded routine counts. Inquiry transcripts detail the questioning of warders and the administrative response that followed.

Richmond Bridewell no longer functions as a prison, yet portions of its structure are documented in historical surveys and maps. Researchers compare architectural plans with surviving descriptions of the corridor where the cell once stood. No shattered lock or broken hinge survives as a relic. Instead, the trace lies in bound volumes and official memoranda, where a name shifts from secured to missing within the span of a single recorded morning.

Chapter 20
The Break at the River Stockade — Simon Fraser and the Convict Flight in New South Wales

Mist lifted slowly from the river flats of New South Wales in the 1830s, revealing a rough stockade built from timber posts driven into damp earth. The enclosure stood near a penal outstation where convicts labored under armed supervision, clearing land and hauling supplies. Morning light filtered through gum trees, casting long shadows across the yard where men assembled for count. Muskets rested against a guard hut constructed from split logs and corrugated bark.

Simon Fraser's name appeared in colonial court proceedings and penal registers following his conviction and assignment to the river settlement. The stockade's design was simple but deliberate, with upright timbers packed tightly to prevent passage. A single gate controlled entry and exit, secured by iron hardware and overseen by sentries. Beyond the fence stretched thick bushland and winding waterways, offering concealment to those who knew the terrain. In the stillness before full daylight, the boundary between confinement and open country lay only a few paces wide.

A Frontier Outstation Under Guard

In the 1830s, river stockades in New South Wales functioned as remote extensions of the colonial penal system. Court proceedings in Sydney recorded convictions and sentences before men were assigned to inland labor settlements. Penal registers listed names, transport ships, and designated outstations. These sites were constructed quickly from local timber, forming rectangular enclosures reinforced with upright posts and a guarded gate.

The river settlement where Simon Fraser was held relied on routine counts and armed oversight rather than stone walls. Inspection reports from the colonial administration describe limited supplies, a small guard detachment, and convicts assigned to clearing land or tending livestock. Movement beyond the stockade required permission and escort. Dense bush and winding waterways surrounded the camp, creating both obstacle and opportunity in terrain that few officials fully controlled.

Timber Posts and Muster Books

The stockade at the river outstation was built from split hardwood posts cut by convict labor and set deep into the soil. Colonial inspection notes describe the uprights driven close together, with crosspieces nailed along the interior to strengthen the barrier. A single timber gate hung on iron hinges forged in Sydney and secured with a padlock issued through government stores. The guard hut stood just inside the entrance, stocked with muskets and ammunition recorded in supply ledgers.

Simon Fraser's confinement was entered in penal registers that tracked assignments to outlying settlements. Daily muster books listed each man present at sunrise and sunset. Labor details were noted beside names, and any absence required written explanation forwarded to the district magistrate. The system relied on head counts and signed entries as much as on the fence of timber posts enclosing the yard.

Eyes on the Bush Line

Within the stockade, convicts watched the rhythm of guard changes and the shifting shadows cast by gum trees along the fence. Fraser's presence was known to overseers who recorded his labor assignments in the muster book. Some guards noted in their daily entries that certain men lingered near the perimeter during tool return. The bush beyond the posts remained thick and uneven, offering cover but also uncertainty for anyone who ventured into it.

Settlers along the river had learned to read signs of movement in the landscape. Reports of escaped convicts occasionally reached magistrates, prompting caution among nearby homesteads. At the outstation, hesitation shaped both sides. Convicts weighed the risk of starvation or capture in open country against confinement within the fence. Guards, aware of the small detachment assigned to the camp, kept a careful eye on gatherings near the gate and along the shadowed timber line.

Measuring the Distance to the Tree Line

By the mid-1830s, colonial authorities had grown concerned about absconding from remote outstations. Circular orders reminded overseers to secure tools at day's end and keep firearms within reach. The district magistrate required written explanations for any discrepancy in count, and additional patrols were occasionally dispatched along known bush tracks. Pressure from Sydney emphasized that even frontier settlements must maintain strict accountability.

Within the stockade, Fraser observed the pattern of labor details that carried men beyond the gate under escort. Religious gatherings on Sundays provided brief intervals when attention shifted toward prayer and hymn. Conversations during woodcutting or river hauling were kept low, timed against the sound of axes and water. The bush offered routes known to Indigenous guides and seasoned bushrangers alike. Under increasing scrutiny from officials, calculation narrowed to moments when routine thinned at the edge of the clearing.

Through the Gate and Into the Bush

Colonial court proceedings and penal correspondence confirm that Simon Fraser absconded from the river stockade during an outside labor assignment. Muster books record his presence at the morning count and note his absence by evening. Reports state that he moved beyond the line of escort while engaged in woodcutting near the tree line. No gate was forced, and no timber post was broken. The breach occurred at the boundary between supervised labor and open bush.

Later frontier tales attributed dramatic confrontations or daring leaps over the fence. Surviving administrative records, however, describe a quieter departure marked by failure to return with the assigned detail. Overseers reported the loss to the district magistrate, and mounted constables were dispatched along nearby tracks. The penal register was amended to list Fraser as absconded from the outstation rather than confined within its timber enclosure.

Evening Count and Mounted Alarm

At sunset muster, Fraser's name was called and met with silence. The overseer compared the tally against the morning entry and confirmed the discrepancy in the muster book. A written report was sent to the district magistrate, and the small guard detachment prepared horses for pursuit. The rifle assigned to the escort detail was also noted in the supply ledger, increasing the urgency of the search.

Suspicion fell upon the supervising overseer, whose judgment was questioned in correspondence forwarded to Sydney. Fellow convicts were confined to the yard under closer watch, and tool access was restricted pending inquiry. Settlers along the river received warnings to report any unfamiliar movement near their homesteads. Within the stockade, strain deepened as accountability tightened, and every remaining man felt the weight of scrutiny tied to a single unreturned name.

Patrols Along the River Bend

Mounted constables followed tracks leading from the woodcutting site toward the river, as recorded in official dispatches preserved in colonial archives. Reports describe search parties moving along known bushranger routes and questioning stockmen at nearby stations. Notices were posted in district towns describing Fraser and warning settlers against offering aid. Each inquiry was logged and forwarded to the magistrate's office.

Within the outstation, procedures shifted immediately. Convicts assigned to outside labor were kept in closer formation, and escorts were instructed to maintain constant visual contact. Firearms were secured when not in direct use, and additional head counts were conducted before and after tool issue. Correspondence with Sydney requested guidance on preventing further absconding from frontier settlements. Fraser's name remained in the penal register as absent, and silence followed each failed report of his capture.

Reinforced Posts and Recorded Lessons

In the years following Fraser's flight, colonial administrators reviewed procedures at remote outstations. Correspondence between district magistrates and Sydney recommended reinforcing timber fences and increasing the number of armed escorts on outside labor details. Inspection notes from later visits reference tightened gate security and stricter accounting of tools and firearms. Muster books required clearer signatures to confirm responsibility at each count.

Decades later, reports summarizing the transportation system cited absconding from frontier settlements as a persistent challenge. Archival files reveal repeated concern that vast bushland and limited manpower weakened enforcement beyond central prisons. Some escapees were later recaptured, others vanished from the record. Fraser's name remains in penal registers marked absconded, yet surviving documents offer no full account of his path through river and forest, leaving gaps between official lines.

Timber Rot and Faded Ink

Today, little remains of the original river stockade beyond archaeological traces and references in colonial maps. Soil depressions and scattered timber fragments mark where posts once stood along the clearing's edge. Survey reports conducted in the twentieth century identified the approximate boundary of the enclosure through ground disturbance patterns. The surrounding bush has reclaimed much of the open yard once kept under watch.

Fraser's conviction and assignment survive in New South Wales penal registers preserved in state archives. The muster book entry marking his absence remains legible, followed by notations of pursuit and unanswered inquiry. Historians handle these volumes carefully, comparing dispatches and district correspondence to trace the sequence of events. No broken gate is displayed behind glass. Instead, the record of a single evening count rests in ink, where a name stands apart from the column of returned men.

Made in the USA
Monee, IL
04 November 2026

67381301R00095